Fundamental Algorithms in Computational Fluid Dynamics

Scientific Computation

For further volumes:
http://www.springer.com/series/718

Thomas H. Pulliam · David W. Zingg

Fundamental Algorithms in Computational Fluid Dynamics

Springer

Thomas H. Pulliam
NASA Ames Research Center
Mountain View, CA
USA

David W. Zingg
Institute for Aerospace Studies
Universtiy of Toronto
Toronto, ON
Canada

ISSN 1434-8322
ISBN 978-3-319-38156-5 ISBN 978-3-319-05053-9 (eBook)
DOI 10.1007/978-3-319-05053-9
Springer Cham Heidelberg New York Dordrecht London

Printed on acid-free paper

Springer is part of Springer Science+Business Media (www.springer.com)

To Harv Lomax and Joe Steger

Preface

The field of computational fluid dynamics (CFD) has matured substantially over the past 30 years and has proven its worth in numerous areas of science and engineering. Although there are different numerical algorithms that can be used to solve the equations governing fluid flow, the key algorithmic concepts that provide the foundations of CFD are by now well established. The purpose of this book is to give a detailed and comprehensive description of some important and widely used algorithms. While the field of CFD will continue to evolve, the algorithms described in this book will continue to provide a basis for understanding most numerical approaches to the solution of the Euler and Navier-Stokes equations in the foreseeable future.

This book is intended to be used as a textbook, that is, within the context of a course in CFD. It is suitable as either a first or a second course at the senior undergraduate or graduate level. As a result of the popularity of CFD, the number of engineers and scientists using it has greatly increased. Many of these users will not have a graduate-level education in CFD; hence the field is increasingly being covered in the undergraduate curricula. It is important to recognize that even a user (as opposed to a developer) of CFD needs some exposure to the underlying theory of both fluid dynamics and numerical algorithms in order to make intelligent use of CFD. The material in this book is equally appropriate to both users and developers of computational methods for fluid dynamics.

In a sense, this is a sequel to our previous book, *Fundamentals of Computational Fluid Dynamics*, written with Harvard Lomax. Whereas that book deals primarily with simple model equations, this one concentrates on the Euler and Navier-Stokes equations. The two books can be used quite naturally in a two-course sequence.[1] However, the present book can also be used as a first course in CFD, as long as the students have had some previous exposure to basic numerical methods. Chapter 2 provides a concise summary of some of the key ideas in our earlier book. Our emphasis is again on a detailed treatment of specific core topics rather than a comprehensive treatment of the entire field. Moreover, our focus is on mature algorithms as opposed to those currently under development.

[1] They have been used in this manner at the University of Toronto for many years.

Given our emphasis on depth as opposed to breadth, several important topics are deferred. Examples include spatial discretizations on unstructured grids, finite-element and spectral methods, and turbulence models, which are left to authors with greater experience in these particular subjects. The algorithms presented here form the basis of several important flow solvers and have been used for countless computations. While many other worthy algorithms exist, understanding and programming the algorithms primarily emphasized in this book should provide the reader with a basis for understanding virtually any algorithm in use today.

A key feature of this text is the use of examples and programming assignments based on the one-dimensional and quasi-one-dimensional Euler equations. Of course, these equations omit important physics (viscosity, heat conduction, and turbulence) and numerics (factorization, meshing). Nevertheless, a great deal can be learned from implementing and studying the algorithms in this context, and the assignments are feasible within a typical one-term course. In order to derive the full benefit of this text, the reader is encouraged to complete the programming assignments associated with each chapter.

We present this book in the hope that it will contribute to an intelligent and creative approach to the development and application of CFD.

Moffett Field, August 2013
Toronto

Thomas H. Pulliam
David W. Zingg

Contents

Chapter 1
Introduction

1.1 Background

The field of computational fluid dynamics (CFD) is the subset of computational science concerned with the solution of the equations governing fluid flow. Although its birth date cannot be pinpointed precisely, it can be said to have begun in earnest in the 1960s. Not surprisingly, this coincides with the development of practical computers. However, the development of the pertinent theory began much earlier. A 1950 paper by Von Neumann and Richtmyer [1] contains a surprising number of the ideas of modern CFD. Furthermore, names such as Gauss, Richardson, and Courant, all of whom predate computers, crop up regularly in the CFD literature. Nevertheless, the development and application of CFD has paralleled that of computers. It is interesting to note that the concept of CFD was envisioned as soon as computers became a reality. For example, in 1946 Alan Turing remarked of the computer he was developing that it " ... would be well adapted to deal with heat transfer problems, at any rate in solids or in fluids without turbulent motion" [2].

In addition to the emergence of viable computers, a second impetus for CFD comes from the inherent difficulty in obtaining general analytical solutions to the equations governing the flow of a fluid, the nonlinear partial differential equations known as the Navier-Stokes equations. At the core of CFD are algorithms for the numerical solution of these equations. Hence the theory associated with CFD algorithms is closely related to the more general theory of numerical methods for the solution of partial differential equations, appropriately specialized to the Navier-Stokes equations. Moreover, CFD in its broadest sense incorporates many other disciplines, from computational geometry to turbulence modeling.

The scientist or engineer often has a need for a quantitative knowledge of the flow of a fluid, such as the velocity, pressure, density, or temperature of the fluid at various locations in the flow domain, under a specific set of conditions. For the scientist, the purpose may be to gain an understanding of a particular phenomenon, such as turbulence or combustion. The engineer typically uses such information in the design process. In general, there are three means by which a quantitative understanding of

T. H. Pulliam and D. W. Zingg, *Fundamental Algorithms in Computational Fluid Dynamics*, Scientific Computation, DOI: 10.1007/978-3-319-05053-9_1,

the flow field of interest can be found: theory, experiment, and computation. Given that analytical solutions are rarely available, one is often left with two alternatives, experiment and CFD, each of which has its strengths and weaknesses. The primary advantages of CFD are typically cost and speed. The primary advantage of experiments is that they are in principle a true representation of reality. These generalizations are an oversimplification, since many CFD solutions can be time-consuming and require an expensive computer, and some experiments contain significant artifacts. Nevertheless, these characteristics of computation and experiment underly the decision process in choosing one or the other. In the aircraft industry, for example, the relatively low cost of CFD has led to a substantial reduction in the wind-tunnel testing performed when designing a new aircraft. However, final verification of the performance of an aircraft normally involves both wind-tunnel and flight testing.

Computing the solution to a specific flow problem using CFD involves a number of tasks, and for each task there exist numerous different approaches and methodologies. Despite this diversity, several common elements can be identified. Let us consider the following four basic steps involved in developing a useful numerical solution of a flow problem:

1. Problem and geometry specification
2. Mesh generation
3. Numerical solution of governing partial differential equations
4. Post-processing, assessment, and interpretation of results.

In order to discuss each of these tasks in more detail, we will consider a hypothetical problem, that of computing the forces generated on a specified aircraft wing in flight. In practice, one may be interested in knowing these forces for a wide range of flight conditions, but for our purpose here we will restrict our interest to one set of operating conditions.

In order to specify the problem, the operating conditions must be precisely defined. These include the speed of the aircraft, the orientation of the wing, and the state of the fluid through which the wing is flying, i.e. its pressure, density, and temperature. This information permits the calculation of key non-dimensional parameters such as the Reynolds number, Mach number, and Knudsen number. In addition, before embarking on a CFD adventure, one should have some qualitative idea of the answers to the following questions: How soon is the solution needed? What level of accuracy is needed in the forces? In other words, what level of error can be tolerated?

At this stage, based on the information described in the previous paragraph, several decisions must be made that will determine the success or failure of the venture. The following are examples of questions that must be asked: Is this a continuum flow? Will the flow be laminar or turbulent? If the latter, what is known about the onset of turbulence, the location of the transition from laminar to turbulent flow? Can compressibility effects be neglected? The answers to questions like these are needed to address the following critical question:

What governing equations will suffice to describe the expected flow phenomena to the desired level of accuracy?

For laminar flows, the correct answer is often the Navier-Stokes equations, as long as the continuum hypothesis holds, which depends on the Knudsen number. Nevertheless there are further questions to be answered: What is the equation of state of the fluid? Is the fluid Newtonian? If not, how are the viscous stresses defined? How does the viscosity vary with temperature? Are the expected flow phenomena time dependent?

When the flow is turbulent, the situation is much more complicated. The Navier-Stokes equations remain the appropriate governing equations. However, the physical time and space scales associated with high Reynolds number turbulent flows, especially wall-bounded flows, are usually much smaller than the scales associated with the geometry and flow speed. As a consequence, the numerical solution of such turbulent flows is extremely demanding computationally. Therefore, a hierarchy of approaches has been developed for tackling turbulent flow problems, ranging from a complete resolution of all relevant scales (known as direct numerical simulation or DNS [3]) to Reynolds averaging (or Favre averaging) in which the equations are time averaged and the resulting so-called Reynolds stresses are modeled. The time-averaged equations are known as the *Reynolds-averaged Navier-Stokes* (or RANS) equations. The models used for the Reynolds stresses are known as *turbulence models* and can be a significant source of error. In between the DNS and RANS approaches lie intermediate, hybrid approaches such as large-eddy simulation (LES) [4] or detached-eddy simulation (DES) [5]. These are intermediate in terms of both accuracy and computing cost.

The purpose of the above discussion is to demonstrate that a deep understanding of fluid dynamics is needed in order to properly formulate a problem for numerical solution. Next, we should discuss geometry specification and mesh generation, but before we can do that we need to define what we mean by a mesh, and we need at least a qualitative understanding of the errors that occur in CFD computations.

The methods we will describe in this book rely on a mesh, or grid–we will use the two terms interchangeably. A mesh is a collection of points that span the flow domain and are connected in some manner. There are two perspectives from which one may view the concept of a mesh. From a finite-difference perspective, the mesh supplies the points at which the solution is approximated, and the connectivity identifies the neighboring points to be used in constructing the finite-difference approximations. From a finite-volume perspective, the purpose of the mesh is to divide the flow domain into a large number of contiguous subdomains, or cells. Therefore, in two dimensions, the lines connecting the grid nodes are the edges of polygonal cells. In three dimensions, they are the edges of polyhedral cells.

The errors that occur in a computation of a fluid flow can be classified as numerical errors or physical model errors. Unless a suitable mesh is chosen for a specific application, numerical errors can be very large. They are typically reduced by adding additional mesh points to the flow domain. We call this *mesh refinement* and we describe the mesh with the added points as having an increased *density*. In principle, mesh dependent numerical error can be reduced to an arbitrarily small level by refining the mesh. With finite precision arithmetic, round-off error prevents the error from being reduced below some lower bound. In practice, this lower bound is rarely approached due to the high computing costs associated with such a highly refined mesh. This has two important implications. First, it means that a certain degree of error is generally accepted. Thus it is important to have an understanding of this error and a means of measuring and controlling it. Second, it means that the properties of the mesh have a significant effect on the accuracy as well as the computational expense of the solution. Consequently a good understanding of both the flow and the algorithm is needed to generate an effective mesh. This motivates the idea of *solution-adaptive meshing*, in which the mesh is determined automatically as part of the solution process.

Numerical errors can be further subdivided into those that are dependent on the refinement of the computational mesh and those that are not. An example of the latter is the error arising as a result of performing an external flow computation on a finite computational domain, which implies enforcing boundary conditions at a finite distance from the body that theoretically should be applied an infinite distance from the body. We will not discuss such errors further at this stage except to remind the reader to be aware of them and to take the necessary steps to reduce them to appropriate levels. Another important type of error is round-off error, which results from finite precision arithmetic. For example, when calculating the difference between two numbers whose difference is many orders of magnitude smaller than either of the numbers, many significant digits are lost. Round-off error has implications for development and implementation of numerical algorithms. However, in practical CFD computations, it is rarely the predominant source of error.

Physical model errors are associated with the various models underlying the governing equations. The largest source of physical model error is often the turbulence model used together with the RANS equations, including the prediction of laminar-turbulent transition. However, it is important to be aware of other possible sources of physical model error, such as the perfect gas law and Sutherland's law, especially if they are used near the limits of their applicability. In comparisons with experiments, incorrect specification of boundary conditions related to the incoming flow can also be a source of error.

Physical model error is much more difficult to estimate and control than numerical error. Physical models must be validated through comparison with reliable experimental data. If the comparison is conducted properly, which means that the numerical errors are negligible compared to the physical model errors, the experimental errors are small, and the computation is an accurate representation of the experiment in terms of geometry, flow conditions, etc., then the physical model error for that particular flow is known. Through a number of such comparisons, a quantitative understanding of the physical model error for a range of flows is obtained. This can be used to estimate the physical model error for a given computation of a flow that lies within this range. If a model is applied outside the range of flows for which it has been validated, then the physical model error is not known and may be very large.

We are now ready to return to the tasks required to solve our hypothetical problem, computing the forces generated on an aircraft wing in flight. Having addressed all of the questions described above and thereby chosen a set of governing equations, let us say the compressible RANS equations, that will represent the physics of our problem with sufficient accuracy, we are ready to generate a mesh. Therefore, we must have a suitable representation of the geometry. In the early days of CFD, the geometry was often represented simply by a surface mesh. Even though a more complete geometry representation may have existed, the mesh generator was provided with only the locations of a finite number of points lying on the surface of the geometry. This approach immediately becomes problematic when a more refined mesh is needed, especially in the context of solution-adaptive meshing. In order to add additional mesh points on the surface of the geometry, some sort of interpolation is needed. The interpolation used, which is dependent on the original surface mesh, then becomes the effective geometry representation. Different geometries can result from different initial meshes and different interpolation techniques. It is far preferable to separate the geometry representation from the mesh generation process and to have a complete and comprehensive geometry representation prior to generating the mesh.

The cost of a CFD computation (both processing time and memory) is dependent on the properties of the mesh and generally increases with the total number of nodes in the mesh. The accuracy of the computation is highly dependent on the properties of the mesh and generally improves as the total number of nodes is increased. Thus there is an inherent compromise between cost and accuracy, and it is important that the mesh points be judiciously distributed in a manner that leads to an efficient computation. For many flows, there are regions in the flow domain where the solution varies much more rapidly than in others. As a result, a uniform distribution of mesh nodes is rarely an efficient strategy. Determining the appropriate mesh density and an efficient placement of the mesh nodes requires an understanding of both the flow solution and the algorithm. This creates a dilemma, since the flow solution is not known prior to its computation. However, for many flows, qualitative features of the flow field can be identified a priori. For example, for our computation of the flow around a wing, since the Reynolds number is presumably large, we know that there will exist thin boundary layers near the surface of the wing. In these regions the flow velocity will change rapidly in the direction normal to the surface of the wing and much less rapidly in the directions parallel to the surface. In order to resolve such a flow field efficiently, the mesh should have a small spacing between mesh points normal to the wing surface and a larger spacing in the other two directions. Experience gained with other similar flows can also be used to guide the mesh generation process. Finally, automated solution-adaptive meshing can alleviate much of the difficulty associated with mesh generation.

The nature of the mesh has important implications for the solution algorithm and vice versa; each approach has advantages and disadvantages. A key distinction is between meshes that are fitted to the geometry and those that are not. Body-fitted meshes simplify the treatment of boundaries, while meshes that are not body-fitted, which are often Cartesian, can be easier to generate and can simplify the algorithm away from the boundaries. Body-fitted meshes can be classified as structured or unstructured; these terms are defined in Chap. 4. Different mesh types can also be combined to exploit their individual advantages. The choice of a specific meshing approach is a critical decision that depends on the geometric complexity and flow physics involved in the problem at hand.

Once a mesh has been generated, we are ready to solve the chosen governing equations. There exist many algorithms with various different properties, and the selection or development of an algorithm for a specific application depends on several factors. This we hope to clarify in this book.

After the solution has been computed, post-processing is required. For example, the forces on our wing must be calculated based on the computed flow solution. It is important that the post-processing calculations be performed in a manner that does not add significantly to the error. In addition, some form of error estimation should be performed, including both numerical and physical model error. If we are to make intelligent use of the calculated forces and moments, for example in the design of an aircraft, it is vital that we have a good understanding of the potential errors in those quantities. Moreover, it is worthwhile to investigate the computed flow solution, usually through flow visualization. This step may reveal that some sort of mundane error has occurred, such as an incorrect input, and, more importantly, it can serve as a check on the initial assumptions made in defining the problem and choosing the governing equations. Of course, a solution computed based on the assumption that the fluid is Newtonian will not provide evidence that it is actually non-Newtonian. However, if the solution has unexpected features, these may challenge some assumptions. If a flow field was computed on the assumption of steady flow, and a large region of flow separation is unexpectedly found, it may be worth recomputing the flow in a time dependent manner. Similarly, examination of the flow solution may reveal that the assumed location of laminar-turbulent transition is suspect or that the mesh is inadequately refined in some regions of the domain.

Finally, there are typically various uncertainties in a computation. Some parameters, such as the shape of the geometry, the angle to the flow, and some coefficients related to the fluid properties, are often known only to within some tolerance. It can be important to have a quantitative understanding of the sensitivity of important outputs of a computation, such as the forces generated by an aircraft wing, to variation in these uncertain input parameters. If the uncertainty in the inputs can be bounded or described in terms of a probability density function, this information can aid in finding a bound on or a probability density function for the outputs.

The goal of the CFD user is to generate a solution that is useful, trustworthy, and accurate; the goal of the CFD developer is to make this as likely as possible. The above discussion is intended to demonstrate that a great deal of knowledge and expertise is needed, of fluid dynamics as well as CFD, not only to develop algorithms and models, but also to apply them successfully. The purpose of this book is to provide the foundation needed to achieve the goals of both users and developers of CFD.

1.2 Overview and Roadmap

What are the foundations upon which the field of CFD is based? In other words, what are the basic topics that anyone intending to make use of CFD should understand? Our answer is as follows: finite-difference methods, finite-volume methods, as well as explicit and implicit time-marching methods. In addition, there are two further topics that have become key ingredients of modern CFD: multigrid and high-resolution upwind schemes. These topics provide the foundations of CFD and are sufficiently mature that it is reasonable to cover them in a basic course in CFD.

Most algorithms used to solve the Euler and Navier-Stokes equations deal with the spatial and temporal aspects of the governing equations separately. Therefore, it is tempting to present them separately, and that is how the presentation of fundamentals proceeds in Chap. 2. However, we choose instead to present two complete algorithms, including both spatial and temporal discretization. One advantage of this approach is that it permits the reader to begin programming earlier. Having learned a complete algorithm in Chap. 4, the reader can immediately program it; this solidifies the understanding of the concepts and provides an opportunity to investigate the behavior of the algorithm. Moreover, we present the two specific algorithms in detail, rather than covering a broader range of different algorithms. This in-depth treatment of these two algorithms provides the reader with a strong basis for understanding other methodologies.

The fundamentals of CFD are covered in Chap. 2. This chapter summarizes much of the material in our previous book and can be omitted by those readers familiar with that book. It introduces the basic concepts of finite-difference methods, the semi-discrete approach, finite-volume methods, time-marching methods, stability analysis, and numerical dissipation in the context of two simple model equations, the linear convection equation and the diffusion equation. The approach is unified and general and provides the background needed to understand the subsequent chapters.

The Euler and Navier-Stokes equations are presented without derivation in Chap. 3 in a form suitable for numerical solution. They are given in both the partial differential equation form solved by finite-difference methods and the integral form solved by finite-volume methods. In addition this chapter introduces the quasi-one-dimensional Euler equations that form the basis of most of the programming assignments. This chapter's exercises require the development of the exact solutions to several one-dimensional problems to be used as a benchmark for the numerical solutions to be developed in the exercises of subsequent chapters.

Chapter 4 presents finite-difference methods and the implicit approximate-factorization algorithm. This classical algorithm forms the basis for many flow solvers, including the widely used NASA codes OVERFLOW [6] and CFL3D [7]. In addition, the generalized curvilinear coordinate transformation, artificial dissipation, and boundary conditions are covered. The exercises in this chapter provide an opportunity to write an implicit finite-difference solver and to apply it to some steady and unsteady problems. Some expected solutions and convergence histories are presented, so the reader can be certain that the algorithm has been properly understood and coded.

Chapter 5 presents a finite-volume method combined with explicit multi-stage time marching and multigrid. This classical algorithm was pioneered by Antony Jameson and his colleagues in various codes designated FLOxx [8–11] and is used in the NASA code TLNS3D [12]. The combination of explicit multi-stage time-marching and multigrid is particularly popular and is used in the NASA code CART3D [13], for example. This chapter's exercises involve coding multi-stage time marching and multigrid. Again, the chapter includes plots of expected convergence behaviour to provide the reader with a reference for comparison.

Finally, Chap. 6 provides an introduction to high-resolution upwind schemes. These have been developed in order to improve the robustness and accuracy of numerical methods for the Euler and Navier-Stokes equations by maintaining some specific physical properties of the solution. For example, if a quantity, such as pressure, is physically required to be positive, then the introduction of a negative pressure as a result of numerical errors could cause significant problems. Calculation of the speed of sound for a perfect gas requires the square root of the pressure or temperature, presupposing that the pressure and temperature are nonnegative. High-resolution schemes are designed to prevent such unphysical occurrences, while maintaining accuracy, and are particularly relevant to flows with shock waves. As a result of their robustness, high-resolution upwind schemes have become quite prevalent in CFD for compressible flows. This chapter presents Godunov's method, which has been influential in the development of high-resolution upwind schemes. The popular approximate Riemann solver of Roe is also described. This leads into an introduction to the principles underlying high-resolution schemes and the presentation of some simple high-resolution upwind schemes with flux limiters. The exercise requires the programming of such a scheme to solve a shock-tube problem.

References

1. Von Neumann, J., and Richtmyer, R.D.: A method for the numerical calculation of hydrodynamic shocks. J. Appl. Phys. **21**, 232–237 (1950)
2. Hodges, A.: Alan Turing: The Enigma. Walker & Co., New York (2000)
3. Moin, P., Mahesh, K.: Direct numerical simulation: a tool in turbulence research. Annu. Rev. Fluid Mech. **30**(1), 539–578 (1998)
4. Sagaut, P.: Large Eddy Simulation for Turbulent Flows. Springer, Berlin (2005)
5. Spalart, P.R.: Strategies for turbulence modelling and simulations. Int. J. Heat Fluid Flow **21**(3), 252–263 (2000)
6. Jespersen, D.C., Pulliam, T.H., Buning, P.G.: Recent enhancements to OVERFLOW (Navier-Stokes code). AIAA Paper 97–0644 (1997)
7. Rumsey, C., Biedron, R., Thomas, J.: CFL3D: its history and some recent applications. NASA TM-112861 (1997)
8. Baker, T.J., Jameson, A., Schmidt, W.: A family of fast and robust Euler codes. Princeton University report MAE 1652 (1984)
9. Jameson, A.: Multigrid algorithms for compressible flow calculations. In: Proceedings of the 2nd European Conference on Multigrid Methods, Lecture Notes in Mathematics 1228. Springer, Heidelberg (1986)
10. Jameson, A., Baker, T.J.: Multigrid solution of the Euler equations for aircraft configurations. AIAA Paper 84–0093 (1984)
11. Swanson, R.C., Turkel, E.: Multistage schemes with multigrid for Euler and Navier-Stokes equations. NASA TP 3631 (1997)
12. Vatsa, V., Wedan, B.: Development of a multigrid code for 3-D Navier-Stokes equations and its application to a grid-refinement study. Comput. Fluids **18**(4), 391–403 (1990)
13. Aftosmis, M.J., Berger, M., Adomavicius, G.: A parallel multilevel method for adaptively refined Cartesian grids with embedded boundaries. AIAA Paper 2000–808 (2000)

Chapter 2
Fundamentals

Before attempting to develop and apply numerical algorithms for the Euler and Navier-Stokes equations it is worthwhile to learn as much as possible by studying the behaviour of such methods when applied to simpler model equations. In this chapter, we will do just that, using two model equations which are linear, scalar partial differential equations (PDEs) that represent physical phenomena relevant to fluid dynamics. This chapter provides a concise summary of our earlier book *Fundamentals of Fluid Dynamics* [1], to which the reader is referred for further details.

2.1 Model Equations

2.1.1 The Linear Convection Equation

The linear convection equation provides a simple model for convection and wave propagation phenomena. It is given by

$$\frac{\partial u}{\partial t} + a\frac{\partial u}{\partial x} = 0, \tag{2.1}$$

where $u(x, t)$ is a scalar quantity propagating with speed a, a real constant which may be positive or negative. In the absence of boundaries, for example on an infinite domain, an initial waveform retains its shape as it propagates in the direction of increasing x if a is positive and in the direction of decreasing x if a is negative. Despite its simplicity, the linear convection equation provides a stiff test for a numerical method, as it is difficult to preserve the initial waveform when it is propagated over long distances.

The linear convection equation is a good model equation in the development of numerical algorithms for the Euler equations, which include both convection and wave propagation phenomena. The one-dimensional Euler equations can be

T. H. Pulliam and D. W. Zingg, *Fundamental Algorithms in Computational Fluid Dynamics*, Scientific Computation, DOI: 10.1007/978-3-319-05053-9_2,

diagonalized so that they can be written as three equations in the form of the linear convection equation, although they of course remain nonlinear and coupled. The quantities propagating are known as *Riemann invariants*, and the speeds at which they propagate are the fluid velocity, the fluid velocity plus the speed of sound, and the fluid velocity minus the speed of sound. If the fluid velocity is positive but less than the speed of sound, i.e. the flow is subsonic, then the first two wave speeds will be positive, and the third will be negative. When using the linear convection equation as a model equation for the Euler equations, one must therefore ensure that wave speeds of arbitrary sign are considered.

If one considers a finite domain, say $0 \leq x \leq 2\pi$, then boundary conditions are required. The most natural boundary conditions are inflow-outflow conditions, which depend on the sign of a. If a is positive, then $x = 0$ is the inflow boundary, and $x = 2\pi$ is the outflow boundary. If a is negative, these roles are reversed. In both cases, $u(t)$ must be specified at the inflow boundary, but no boundary condition can be specified at the outflow boundary.

An alternative specification of boundary conditions, known as periodic boundary conditions, can be convenient for our purpose here. With periodic boundary conditions, a waveform leaving one end of the domain reenters at the other end. The domain can be visualized as a circle, and the waveform simply propagates repeatedly around the circle. This essentially eliminates any boundary information from entering the solution, which is thus determined solely by the initial condition. The use of periodic boundary conditions also permits numerical experiments with arbitrarily long propagation distances, independent of the size of the domain. Each time the initial waveform travels through the entire domain, it should return unaltered to the initial condition.

2.1.2 The Diffusion Equation

Diffusion caused by molecular motion in a continuum fluid is another important physical phenomenon in fluid dynamics. A simple linear model equation for a diffusive process is

$$\frac{\partial u}{\partial t} = \nu \frac{\partial^2 u}{\partial x^2}, \tag{2.2}$$

where ν is a positive real constant. For example, with u representing the temperature, this parabolic PDE governs the diffusion of heat in one dimension. Boundary conditions can be periodic, Dirichlet (specified u), Neumann (specified $\partial u/\partial x$), or mixed Dirichlet/Neumann. In studying numerical algorithms, it can be useful to introduce a source term into the diffusion equation as follows:

$$\frac{\partial u}{\partial t} = \nu \left[\frac{\partial^2 u}{\partial x^2} - g(x) \right]. \tag{2.3}$$

In this case, the equation has a steady-state solution that satisfies

$$\frac{\partial^2 u}{\partial x^2} - g(x) = 0. \tag{2.4}$$

2.2 Finite-Difference Methods

2.2.1 Basic Concepts: Taylor Series

We observe that the two model equations contain a number of derivative terms in space and time. In a finite-difference method, a spatial derivative at a given point in space is approximated using values of u at nearby points in space. Similarly, a temporal derivative at a specific point in time is approximated using values of u at different values of time. This is facilitated by a grid or mesh, as shown in Fig. 2.1, where the values of x at the grid points are given by x_j, and the values of t are given by t_n. Hence j is known as the spatial index and n as the temporal index. For the present exposition, we will consider equispaced meshes, and hence

$$x = x_j = j\Delta x \tag{2.5}$$

$$t = t_n = n\Delta t = nh, \tag{2.6}$$

where Δx is the spacing in x, and Δt the spacing in t, as shown in Fig. 2.1. Note that $h = \Delta t$ throughout.

Let us consider a spatial derivative initially. Assuming that a function $u(x, t)$ is known only at discrete values of x, how can one accurately approximate partial derivatives such as

$$\frac{\partial u}{\partial x} \quad \text{or} \quad \frac{\partial^2 u}{\partial x^2} \quad ? \tag{2.7}$$

From the definition of a derivative, or a simple geometric argument related to the tangent to a curve, one can easily postulate the following approximations for a first derivative:

$$\left(\frac{\partial u}{\partial x}\right)_j \approx \frac{u_{j+1} - u_j}{\Delta x} \quad \text{or} \quad \left(\frac{\partial u}{\partial x}\right)_j \approx \frac{u_j - u_{j-1}}{\Delta x}, \tag{2.8}$$

known respectively as a forward and a backward difference approximation. It is clear that these can provide accurate approximations if Δx is sufficiently small and that a suitable choice of Δx will depend on the properties of the function. A particularly astute reader may even postulate the centered difference approximation given by

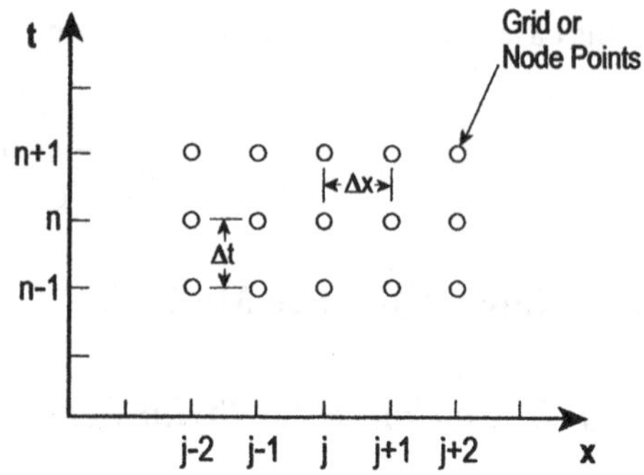

Fig. 2.1 Space–time grid arrangement

$$\left(\frac{\partial u}{\partial x}\right)_j \approx \frac{u_{j+1} - u_{j-1}}{2\Delta x}, \tag{2.9}$$

which can be justified by a geometric argument or by fitting a parabola to the three points u_{j-1}, u_j, u_{j+1} and determining the first derivative of the parabola at x_j. Finding an approximation to a second derivative is only slightly less intuitive, as one can apply a first-derivative approximation twice or determine the second derivative of the unique parabola that goes through the three points to obtain the following approximation to a second derivative:

$$\left(\frac{\partial^2 u}{\partial x^2}\right)_j \approx \frac{u_{j+1} - 2u_j + u_{j-1}}{\Delta x^2}. \tag{2.10}$$

The above intuitive approach is limited, providing no information about the accuracy of these approximations. A deeper understanding of finite-difference approximations and a general approach to deriving them can be obtained using Taylor series. Consider the following expansion of $u(x + k\Delta x) = u(j\Delta x + k\Delta x) = u_{j+k}$ about x_j, where we assume that all of the derivatives exist:

$$\begin{aligned} u_{j+k} = u_j + (k\Delta x)\left(\frac{\partial u}{\partial x}\right)_j + \frac{1}{2}(k\Delta x)^2\left(\frac{\partial^2 u}{\partial x^2}\right)_j + \dots \\ + \frac{1}{n!}(k\Delta x)^n\left(\frac{\partial^n u}{\partial x^n}\right)_j + \dots . \end{aligned} \tag{2.11}$$

For example, substituting $k = \pm 1$ into the above expression gives the Taylor series expansions for $u_{j\pm 1}$:

$$\begin{aligned} u_{j\pm 1} = u_j \pm (\Delta x)\left(\frac{\partial u}{\partial x}\right)_j + \frac{1}{2}(\Delta x)^2\left(\frac{\partial^2 u}{\partial x^2}\right)_j \pm \frac{1}{6}(\Delta x)^3\left(\frac{\partial^3 u}{\partial x^3}\right)_j \\ + \frac{1}{24}(\Delta x)^4\left(\frac{\partial^4 u}{\partial x^4}\right)_j \pm \dots . \end{aligned} \tag{2.12}$$

Subtracting u_j from the Taylor series expansion for u_{j+1} and dividing by Δx gives

$$\frac{u_{j+1} - u_j}{\Delta x} = \left(\frac{\partial u}{\partial x}\right)_j + \frac{1}{2}(\Delta x)\left(\frac{\partial^2 u}{\partial x^2}\right)_j + \dots. \tag{2.13}$$

This shows that the forward difference approximation given in (2.8) is a reasonable approximation for $\left(\frac{\partial u}{\partial x}\right)_j$ as long as Δx is small relative to some pertinent length scale. Moreover, in the limit as $\Delta x \to 0$, the leading term in the error is proportional to Δx. The *order of accuracy* of an approximation is given by the exponent of Δx in the leading error term, i.e. the lowest exponent of Δx in the error. Hence the finite-difference approximation given in (2.13) is a *first-order approximation* to a first derivative. If the mesh spacing Δx is reduced by a factor of two, the leading error term in a first-order approximation will also be reduced by a factor of two.

Similarly, subtracting the Taylor series expansion for u_{j-1} from that for u_{j+1} and dividing by $2\Delta x$ gives

$$\frac{u_{j+1} - u_{j-1}}{2\Delta x} = \left(\frac{\partial u}{\partial x}\right)_j + \frac{1}{6}\Delta x^2\left(\frac{\partial^3 u}{\partial x^3}\right)_j + \frac{1}{120}\Delta x^4\left(\frac{\partial^5 u}{\partial x^5}\right)_j \dots. \tag{2.14}$$

This shows that the centered difference approximation given in (2.9) is second-order accurate. If the mesh spacing Δx is reduced by a factor of two, the leading error term will be reduced by a factor of four. Hence, as Δx is reduced, the second-order approximation rapidly becomes more accurate than the first-order approximation. Using Taylor series expansions, one can demonstrate that the approximation to a second derivative given in (2.10) is also second-order accurate.

Finite-difference formulas can be generalized to arbitrary derivatives and arbitrary orders of accuracy. A Taylor table provides a convenient mechanism for deriving finite-difference operators (see Lomax et al. [1]). In each case, the derivative at node j is approximated using a linear combination of function values at node j and a specified number of neighbouring nodes, and the Taylor table enables one to find the coefficients that maximize the order of accuracy. For example, centered fourth-order approximations to first and second derivatives are given by

$$\left(\frac{\partial u}{\partial x}\right)_j = \frac{1}{12\Delta x}(u_{j-2} - 8u_{j-1} + 8u_{j+1} - u_{j+2}) + O(\Delta x^4) \tag{2.15}$$

$$\left(\frac{\partial^2 u}{\partial x^2}\right)_j = \frac{1}{12\Delta x^2}(-u_{j-2} + 16u_{j-1} - 30u_j + 16u_{j+1} - u_{j+2}) + O(\Delta x^4). \tag{2.16}$$

Noncentered schemes can also be useful. For example, the following is a second-order backward-difference approximation to a first derivative using data from $j-2$ to j:

$$\left(\frac{\partial u}{\partial x}\right)_j = \frac{1}{2\Delta x}\left(u_{j-2} - 4u_{j-1} + 3u_j\right) + O(\Delta x^2). \tag{2.17}$$

A *biased* third-order approximation to a first derivative using data from $j-2$ to $j+1$ is given by:

$$\left(\frac{\partial u}{\partial x}\right)_j = \frac{1}{6\Delta x}(u_{j-2} - 6u_{j-1} + 3u_j + 2u_{j+1}) + O(\Delta x^3). \tag{2.18}$$

Finally, finite-difference schemes can be further generalized to include *compact or Padé* schemes that define a linear combination of the derivatives at point j and a specified number of its neighbours as a linear combination of function values at node j and a (possibly different) specified number of neighbours. For example, the operator

$$\left(\frac{\partial u}{\partial x}\right)_{j-1} + 4\left(\frac{\partial u}{\partial x}\right)_j + \left(\frac{\partial u}{\partial x}\right)_{j+1} = \frac{3}{\Delta x}(-u_{j-1} + u_{j+1}) + O(\Delta x^4). \tag{2.19}$$

provides a fourth-order approximation to a first derivative. Compact schemes can also be easily derived using a Taylor table.

2.2.2 The Modified Wavenumber

The leading error term provides a fairly limited understanding of the accuracy of a finite-difference approximation. More detailed information can be obtained through the *modified wavenumber*. We introduce this concept by deriving the modified wavenumber for a second-order centered difference approximation, given by

$$(\delta_x u)_j = \frac{u_{j+1} - u_{j-1}}{2\Delta x}. \tag{2.20}$$

First, consider the exact first derivative of the function $\mathrm{e}^{\mathrm{i}\kappa x}$:

$$\frac{\partial \mathrm{e}^{\mathrm{i}\kappa x}}{\partial x} = \mathrm{i}\kappa \mathrm{e}^{\mathrm{i}\kappa x}. \tag{2.21}$$

Applying the operator given in (2.20) to $u_j = \mathrm{e}^{\mathrm{i}\kappa x_j}$, where $x_j = j\Delta x$, we get

$$\begin{aligned}(\delta_x u)_j &= \frac{\mathrm{e}^{\mathrm{i}\kappa\Delta x(j+1)} - \mathrm{e}^{\mathrm{i}\kappa\Delta x(j-1)}}{2\Delta x}\\ &= \frac{(\mathrm{e}^{\mathrm{i}\kappa\Delta x} - \mathrm{e}^{-\mathrm{i}\kappa\Delta x})\mathrm{e}^{\mathrm{i}\kappa x_j}}{2\Delta x}\end{aligned}$$

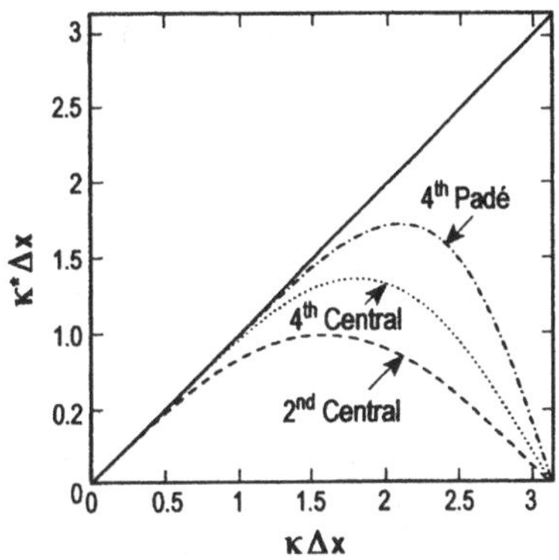

Fig. 2.2 Modified wavenumber for various schemes

$$
\begin{aligned}
&= \frac{1}{2\Delta x}[(\cos \kappa\Delta x + \mathrm{i}\sin \kappa\Delta x) - (\cos \kappa\Delta x - \mathrm{i}\sin \kappa\Delta x)]\mathrm{e}^{\mathrm{i}\kappa x_j} \\
&= \mathrm{i}\frac{\sin \kappa\Delta x}{\Delta x}\mathrm{e}^{\mathrm{i}\kappa x_j} \\
&= \mathrm{i}\kappa^* \mathrm{e}^{\mathrm{i}\kappa x_j},
\end{aligned} \tag{2.22}
$$

where κ^* is the modified wavenumber. The modified wavenumber is so named because it appears where the wavenumber, κ, appears in the exact expression (2.21). Thus the degree to which the modified wavenumber approximates the actual wavenumber is a measure of the accuracy of the approximation.

For the second-order centered difference operator the modified wavenumber is given by

$$
\kappa^* = \frac{\sin \kappa\Delta x}{\Delta x}. \tag{2.23}
$$

Equation (2.23) is plotted in Fig. 2.2, along with similar relations for the standard fourth-order centered difference scheme and the fourth-order Padé scheme. The expression for the modified wavenumber provides the accuracy with which a given wavenumber component of the solution is resolved for the entire wavenumber range available in a mesh of a given size, $0 \le \kappa\Delta x \le \pi$. The value of $\kappa\Delta x$ can be related to the mesh resolution through the notion of points-per-wavelength, which is the number of grid cells per wavelength (PPW) with which a given wavenumber component of the solution is resolved, through the relation $PPW = 2\pi/\kappa\Delta x$. For example, a value of $\kappa\Delta x$ equal to $\pi/4$ corresponds to 8 points-per-wavelength, and Fig. 2.2 shows that κ^* for a second-order centered difference scheme already differs significantly from κ at this grid resolution. Hence a simulation performed with this mesh density relative to the spectral content of the function will contain substantial numerical error.

For centered difference approximations, the modified wavenumber is purely real, but in the general case it can include an imaginary component as well. Any finite-difference operator can be split into an antisymmetric and a symmetric part. For example, the operator given in (2.18) can be divided as follows:

$$\begin{aligned}(\delta_x u)_j &= \frac{1}{6\Delta x}(u_{j-2} - 6u_{j-1} + 3u_j + 2u_{j+1}) \\ &= \frac{1}{12\Delta x}[(u_{j-2} - 8u_{j-1} + 8u_{j+1} - u_{j+2}) \\ &\quad + (u_{j-2} - 4u_{j-1} + 6u_j - 4u_{j+1} + u_{j+2})]. \end{aligned} \tag{2.24}$$

The antisymmetric component determines the real part of the modified wavenumber, while the imaginary part stems from the symmetric component of the difference operator. Centered difference schemes are antisymmetric; the symmetric component is zero and hence so is the imaginary component of the modified wavenumber. In the context of the linear convection equation, one can show that a numerical error in the phase speed is associated with the real part of the modified wavenumber, while an error in the amplitude of the solution is associated with the imaginary part. Thus the antisymmetric portion of the spatial difference operator determines the error in speed and the symmetric portion the error in amplitude. Note that centered schemes produce no amplitude error. Since the numerical error in the phase speed is dependent on the wavenumber, this introduces numerical dispersion, and hence phase speed error is often referred to as *dispersive error*. Similarly, amplitude error is often referred to as *dissipative* error.

2.3 The Semi-Discrete Approach

Based on the discussion in the previous section, one can see that it is possible to replace both the spatial and temporal derivatives in a PDE by finite-difference expressions and thereby reduce the PDE to a system of algebraic equations that can be solved by a computer. For various reasons it can be advantageous to consider the discretization of space and time separately. We first discretize in space to reduce the PDE to a system of ordinary differential equations (ODEs) in the general form

$$\frac{d\vec{u}}{dt} = \vec{F}(\vec{u}, t), \tag{2.25}$$

and then apply a time-marching method to reduce the system of ODEs to a system of algebraic equations in order to solve them. This is referred to as the *semi-discrete approach*, and the intermediate ODE form in which the spatial derivatives have been discretized but the temporal derivatives have not is known as the *semi-discrete form*. It is important to realize that some numerical algorithms for PDEs discretize time and space simultaneously and consequently have no intermediate semi-discrete form. However, many of the most widely used algorithms and all of those considered in subsequent chapters involve a separate and distinct discretization in time and space.

In the semi-discrete approach, one separates the spatial discretization step that reduces the PDE to a system of ODEs from the time-marching step that numerically solves the ODE system. By doing so, we can get a clear understanding of the impact on accuracy and stability of the spatial discretization and the time-marching method individually. This approach also enables us to take advantage of the theory associated with numerical methods for ODEs. Before we can present the semi-discrete ODE form for our model equations, we need an understanding of *matrix difference operators*.

2.3.1 Matrix Difference Operators

Consider the relation

$$(\delta_{xx}u)_j = \frac{1}{\Delta x^2}\left(u_{j+1} - 2u_j + u_{j-1}\right), \tag{2.26}$$

which is a point difference approximation to a second derivative. Now let us derive a *matrix* operator representation for the same approximation. Consider the mesh spanning the domain $0 \le x \le \pi$ with four interior points and boundary points labelled a and b shown below.

$$\begin{array}{ccccccc} & a & 1 & 2 & 3 & 4 & b \\ x = & 0 & - & - & - & - & \pi \\ j = & & 1 & \cdot & \cdot & M & \end{array}$$

Mesh with four interior points. $\Delta x = \pi/(M+1)$

Now impose Dirichlet boundary conditions, $u(0) = u_a,\ u(\pi) = u_b$ and use the centered difference approximation given by (2.26) at every point in the mesh. We arrive at the four equations:

$$\begin{aligned} (\delta_{xx}u)_1 &= \frac{1}{\Delta x^2}(u_a - 2u_1 + u_2) \\ (\delta_{xx}u)_2 &= \frac{1}{\Delta x^2}(u_1 - 2u_2 + u_3) \\ (\delta_{xx}u)_3 &= \frac{1}{\Delta x^2}(u_2 - 2u_3 + u_4) \\ (\delta_{xx}u)_4 &= \frac{1}{\Delta x^2}(u_3 - 2u_4 + u_b). \end{aligned} \tag{2.27}$$

Introducing

$$\vec{u} = \begin{bmatrix} u_1 \\ u_2 \\ u_3 \\ u_4 \end{bmatrix}, \quad \left(\vec{bc}\right) = \frac{1}{\Delta x^2} \begin{bmatrix} u_a \\ 0 \\ 0 \\ u_b \end{bmatrix} \tag{2.28}$$

and

$$A = \frac{1}{\Delta x^2} \begin{bmatrix} -2 & 1 & & \\ 1 & -2 & 1 & \\ & 1 & -2 & 1 \\ & & 1 & -2 \end{bmatrix}, \tag{2.29}$$

we can rewrite (2.27) as

$$\delta_{xx}\vec{u} = A\vec{u} + \left(\vec{bc}\right). \tag{2.30}$$

This example illustrates a matrix difference operator. Each line of a matrix difference operator is based on a point difference operator, but the point operators used from line to line are not necessarily the same. For example, boundary conditions may dictate that the lines at or near the bottom or top of the matrix be modified. In the extreme case of the matrix difference operator representing a spectral method, none of the lines is the same. The matrix operators representing the three-point central-difference approximations for a first and second derivative with Dirichlet boundary conditions on a four-point mesh are

$$\delta_x = \frac{1}{2\Delta x} \begin{bmatrix} 0 & 1 & & \\ -1 & 0 & 1 & \\ & -1 & 0 & 1 \\ & & -1 & 0 \end{bmatrix}, \quad \delta_{xx} = \frac{1}{\Delta x^2} \begin{bmatrix} -2 & 1 & & \\ 1 & -2 & 1 & \\ & 1 & -2 & 1 \\ & & 1 & -2 \end{bmatrix}. \tag{2.31}$$

Each of these matrix difference operators is a square matrix with elements that are all zeros except for those along bands which are clustered around the central diagonal. We call such a matrix a *banded matrix* and introduce the notation

$$B(M : a, b, c) = \begin{bmatrix} b & c & & & \\ a & b & c & & \\ & & \ddots & & \\ & & a & b & c \\ & & & a & b \end{bmatrix} \begin{matrix} 1 \\ \\ \vdots \\ \\ M \end{matrix}, \tag{2.32}$$

where the matrix dimensions are $M \times M$. Use of M in the argument is optional, and the illustration is given for a simple *tridiagonal* matrix although any number of

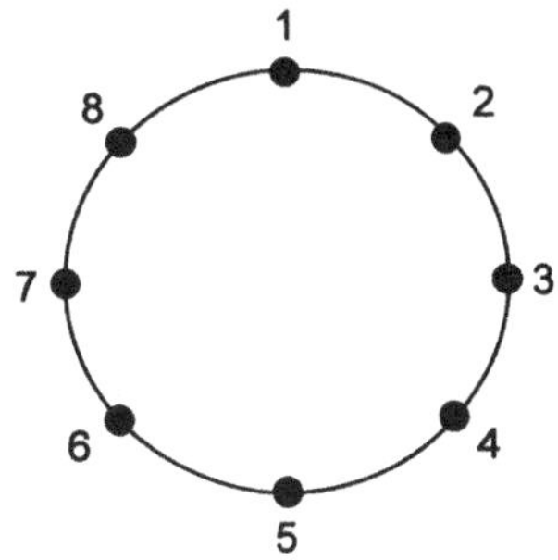

Fig. 2.3 Eight points on a circular mesh

bands is a possibility. A tridiagonal matrix without constants along the bands can be expressed as $B(\vec{a}, \vec{b}, \vec{c})$. The arguments for a banded matrix are always odd in number, and the central one *always* refers to the central diagonal.

If the boundary conditions are *periodic*, the form of the matrix operator changes. Consider the eight-point periodic mesh spanning the domain $0 \leq x \leq 2\pi$ shown below. This can either be presented on a linear mesh with repeated entries, or more suggestively on a circular mesh as in Fig. 2.3. When the mesh is laid out on the perimeter of a circle, it does not matter where the numbering starts, as long as it "ends" at the point just preceding its starting location.

$$
\begin{array}{llllllllllllll}
\cdots & 7 & 8 & 1 & 2 & 3 & 4 & 5 & 6 & 7 & 8 & 1 & 2 & \cdots \\
x = & - & - & 0 & - & - & - & - & - & - & - & 2\pi & - & \\
j = & & & 0 & 1 & \cdot & \cdot & \cdot & \cdot & \cdot & \cdot & M & &
\end{array}
$$

Eight points on a linear periodic mesh. $\Delta x = 2\pi/M$

The matrix that represents differencing schemes for scalar equations on a periodic mesh is referred to as a *periodic* matrix. A special subset of a periodic matrix is a *circulant* matrix, formed when the elements along the various bands are constant. Each row of a circulant matrix is shifted one element to the right of the one above it. The special case of a tridiagonal circulant matrix is given by

$$
B_{\mathrm{p}}(M : a, b, c) = \begin{bmatrix} b & c & & & a \\ a & b & c & & \\ & & \ddots & & \\ & & a & b & c \\ c & & & a & b \end{bmatrix} \begin{matrix} 1 \\ \\ \vdots \\ \\ M \end{matrix} . \tag{2.33}
$$

When the standard three-point central-differencing approximation for a first derivative (see (2.31)) is used with periodic boundary conditions, it takes the form

$$(\delta_x)_\mathrm{p} = \frac{1}{2\Delta x}\begin{bmatrix} 0 & 1 & & -1 \\ -1 & 0 & 1 & \\ & -1 & 0 & 1 \\ 1 & & -1 & 0 \end{bmatrix} = \frac{1}{2\Delta x}B_\mathrm{p}(-1, 0, 1).$$

Notice that there is no boundary condition vector since this information is all interior to the matrix itself.

2.3.2 Reduction of PDEs to ODEs

Now that we have the concept of matrix difference operators, we can proceed to apply a spatial discretization to reduce PDEs to ODEs. First let us consider the model PDEs for diffusion and periodic convection described in Sect. 2.1. In these simple cases, we can approximate the space derivatives with difference operators and express the resulting ODEs with a matrix formulation. This is a simple and natural formulation when the ODEs are linear.

Model ODE for Diffusion. For example, using the three-point central-differencing scheme to represent the second derivative in the scalar PDE governing diffusion leads to the following ODE diffusion model:

$$\frac{\mathrm{d}\vec{u}}{\mathrm{d}t} = \frac{\nu}{\Delta x^2}B(1, -2, 1)\vec{u} + (\vec{bc}) \tag{2.34}$$

with Dirichlet boundary conditions folded into the $(\vec{bc})$ vector.

Model ODE for Periodic Convection. For the linear convection equation with periodic boundary conditions, the 3-point central-differencing approximation produces the ODE model given by

$$\frac{\mathrm{d}\vec{u}}{\mathrm{d}t} = -\frac{a}{2\Delta x}B_\mathrm{p}(-1, 0, 1)\vec{u}, \tag{2.35}$$

where the boundary condition vector is absent because the flow is periodic.

Equations (2.34) and (2.35) are the model ODEs for diffusion and periodic convection of a scalar in one dimension. They are linear with coefficient matrices which are independent of x and t.

The Generic Matrix Form. The generic matrix form of a semi-discrete approximation is expressed by the equation

$$\frac{\mathrm{d}\vec{u}}{\mathrm{d}t} = A\vec{u} - \vec{f}(t). \tag{2.36}$$

Note that the elements in the matrix A depend upon both the PDE and the type of differencing scheme chosen for the space terms. The vector $\vec{f}(t)$ is usually determined by the boundary conditions and possibly source terms. In general, even the Euler and

Navier–Stokes equations can be expressed in the form of (2.36). In such cases the equations are nonlinear, that is, the elements of A depend on the solution $\vec{u}$ and are usually derived by finding the Jacobian of a flux vector. Although the equations are nonlinear, linear analysis leads to diagnostics that are surprisingly accurate when used to evaluate many aspects of numerical methods as they apply to the Euler and Navier–Stokes equations.

2.3.3 Exact Solutions of Linear ODEs

In order to advance (2.25) in time, the system of ODEs must be integrated using a time-marching method. In order to analyze time-marching methods, we will make use of exact solutions of coupled systems of ODEs, which exist under certain conditions. The ODEs represented by (2.25) are said to be *linear* if F is linearly dependent on u (i.e. if $\partial F/\partial u = A$, where A is independent of u). As we have already pointed out, when the ODEs are linear they can be expressed in a matrix notation as (2.36) in which the coefficient matrix, A, is independent of u. If A *does* depend explicitly on t, the general solution *cannot* be written, whereas, if A *does not* depend explicitly on t, the general solution to (2.36) *can* be written. This holds regardless of whether or not the forcing function, $\vec{f}$, depends explicitly on t.

The exact solution of (2.36) can be written in terms of the eigenvalues and eigenvectors of A. This will lead us to a representative scalar equation for use in analyzing time-marching methods. To demonstrate this, let us consider a set of coupled, nonhomogeneous, linear, first-order ODEs with constant coefficients which might have been derived by space differencing a set of PDEs. Represent them by the equation

$$\frac{\mathrm{d}\vec{u}}{\mathrm{d}t} = A\vec{u} - \vec{f}(t). \tag{2.37}$$

Our assumption is that the $M \times M$ matrix A has a complete eigensystem[1] and can thus be transformed by the left and right eigenvector matrices, X^{-1} and X, to a diagonal matrix Λ having diagonal elements which are the eigenvalues of A. Now let us multiply (2.37) from the left by X^{-1} and insert the identity combination $XX^{-1} = I$ between A and $\vec{u}$. There results

$$X^{-1}\frac{\mathrm{d}\vec{u}}{\mathrm{d}t} = X^{-1}AX \cdot X^{-1}\vec{u} - X^{-1}\vec{f}(t). \tag{2.38}$$

Since A is independent of both $\vec{u}$ and t, the elements in X^{-1} and X are also independent of both $\vec{u}$ and t, and (2.38) can be modified to

---

[1] This means that the eigenvectors of A are linearly independent and thus $X^{-1}AX = \Lambda$, where X contains the right eigenvectors of A as its columns, i.e. $X = [\vec{x}_1\ ,\ \vec{x}_2\ \ldots\ ,\vec{x}_M]$, and Λ is a diagonal matrix whose elements are the eigenvalues of A.

$$\frac{\mathrm{d}}{\mathrm{d}t}X^{-1}\vec{u} = \Lambda X^{-1}\vec{u} - X^{-1}\vec{f}(t).$$

Finally, by introducing the new variables $\vec{w}$ and $\vec{g}$ such that

$$\vec{w} = X^{-1}\vec{u}, \quad \vec{g}(t) = X^{-1}\vec{f}(t), \tag{2.39}$$

we reduce (2.37) to a new algebraic form

$$\frac{\mathrm{d}\vec{w}}{\mathrm{d}t} = \Lambda\vec{w} - \vec{g}(t). \tag{2.40}$$

The equations represented by (2.40) are no longer coupled. They can be written line by line as a set of independent, single, first-order equations, thus

$$\begin{aligned} w_1' &= \lambda_1 w_1 - g_1(t) \\ &\vdots \\ w_m' &= \lambda_m w_m - g_m(t) \\ &\vdots \\ w_M' &= \lambda_M w_M - g_M(t). \end{aligned} \tag{2.41}$$

For any given set of $g_m(t)$ each of these equations can be solved separately and then recoupled, using the inverse of the relations given in (2.39):

$$\begin{aligned} \vec{u}(t) &= X\vec{w}(t) \\ &= \sum_{m=1}^{M} w_m(t)\vec{x}_m, \end{aligned} \tag{2.42}$$

where $\vec{x}_m$ is the mth column of X, i.e. the eigenvector corresponding to λ_m.

We next focus on the important subset of (2.36) when neither A nor $\vec{f}$ has any explicit dependence on t. In such a case, the g_m in (2.40) and (2.41) are also time invariant, and the solution to any line in (2.41) is

$$w_m(t) = c_m \mathrm{e}^{\lambda_m t} + \frac{1}{\lambda_m} g_m,$$

where the c_m are constants that depend on the initial conditions. Transforming back to the u-system gives

$$\begin{aligned} \vec{u}(t) &= X\vec{w}(t) \\ &= \sum_{m=1}^{M} w_m(t)\vec{x}_m \end{aligned}$$

$$
\begin{aligned}
&= \sum_{m=1}^{M} c_m e^{\lambda_m t} \vec{x}_m + \sum_{m=1}^{M} \frac{1}{\lambda_m} g_m \vec{x}_m \\
&= \sum_{m=1}^{M} c_m e^{\lambda_m t} \vec{x}_m + X \Lambda^{-1} X^{-1} \vec{f} \\
&= \underbrace{\sum_{m=1}^{M} c_m e^{\lambda_m t} \vec{x}_m}_{\text{transient}} + \underbrace{A^{-1} \vec{f}}_{\text{steady-state}} .
\end{aligned}
\tag{2.43}
$$

Note that the steady-state solution is $A^{-1}\vec{f}$, as might be expected.

The first group of terms on the right side of this equation is referred to classically as the *complementary solution* or the solution of the homogeneous equations. The second group is referred to classically as the *particular solution* or the particular integral. In our application to fluid dynamics, it is more descriptive to refer to these groups as the *transient* and *steady-state* solutions, respectively. An alternative, but entirely equivalent, form of the solution is

$$
\vec{u}(t) = c_1 e^{\lambda_1 t} \vec{x}_1 + \cdots + c_m e^{\lambda_m t} \vec{x}_m + \cdots + c_M e^{\lambda_M t} \vec{x}_M + A^{-1} \vec{f}. \tag{2.44}
$$

2.3.4 Eigenvalue Spectra for Model ODEs

It is instructive to consider the eigenvalue spectra of the ODEs formulated by central differencing the model equations for diffusion (2.34) and periodic convection (2.35). For the model diffusion equation with Dirichlet boundary conditions, the eigenvalues of A are:

$$
\begin{aligned}
\lambda_m &= \frac{\nu}{\Delta x^2} \left[-2 + 2\cos\left(\frac{m\pi}{M+1}\right) \right] \\
&= \frac{-4\nu}{\Delta x^2} \sin^2\left(\frac{m\pi}{2(M+1)}\right), \quad m = 1, 2, \ldots, M.
\end{aligned}
\tag{2.45}
$$

These eigenvalues are all real and negative, consistent with the physics of diffusion.

For periodic convection, one obtains

$$
\begin{aligned}
\lambda_m &= \frac{-\mathrm{i}a}{\Delta x} \sin\left(\frac{2m\pi}{M}\right), \quad m = 0, 1, \ldots, M-1 \\
&= -\mathrm{i}\kappa_m^* a,
\end{aligned}
\tag{2.46}
$$

where

$$\kappa_m^* = \frac{\sin \kappa_m \Delta x}{\Delta x}, \qquad m = 0, 1, \ldots, M-1 \tag{2.47}$$

is the modified wavenumber, $\kappa_m = m$, and $\Delta x = 2\pi/M$. These eigenvalues are all pure imaginary, reflecting the fact that the amplitude of a waveform neither grows nor decays as it convects, a property that is preserved by centered differencing.

2.3.5 A Representative Equation for Studying Time-Marching Methods

We seek to analyze the accuracy and stability of time-marching methods applied to the systems of ODEs resulting from applying a spatial discretization to PDEs such as the Navier-Stokes equations, which take the form:

$$\frac{\mathrm{d}\vec{u}}{\mathrm{d}t} = \vec{F}(\vec{u}, t), \tag{2.48}$$

To simplify matters, we consider the simpler model equations, which lead to ODE forms such as (2.34) and (2.35) that have the generic form:

$$\frac{\mathrm{d}\vec{u}}{\mathrm{d}t} = A\vec{u} - \vec{f}(t), \tag{2.49}$$

where A is independent of u and t. To achieve a further simplification, we exploit the fact that these equations can be decoupled and study time-marching methods as applied to the following scalar ODE:

$$\frac{\mathrm{d}u}{\mathrm{d}t} = \lambda u + a\mathrm{e}^{\mu t}, \tag{2.50}$$

where λ, a, and μ are complex constants. The goal in our analysis is to study typical behavior of general situations, not particular problems. In order to evaluate time-marching methods, the parameters λ, a, and μ must be allowed to take the worst possible combination of values that might occur in the ODE eigensystem. For example, if one is interested in a time-marching method for convection dominated problems, then one should consider imaginary λs. The exact solution of the representative ODE is (for $\mu \neq \lambda$)

$$u(t) = c\,\mathrm{e}^{\lambda t} + \frac{a\mathrm{e}^{\mu t}}{\mu - \lambda}, \tag{2.51}$$

where the constant c is determined from the initial condition.

2.4 Finite-Volume Methods

2.4.1 Basic Concepts

We saw in Sect. 2.3 that a PDE can be reduced to a system of ODEs by discretizing the spatial derivatives using finite-difference approximations. A finite-volume method is an alternative spatial discretization that reduces the integral form of a conservation law to a system of ODEs. Finite-volume methods have become popular in CFD as a result, primarily, of two advantages. First, they ensure that the discretization is conservative, i.e. mass, momentum, and energy are conserved in a discrete sense. While this property can usually be obtained using a finite-difference formulation, it is obtained naturally from a finite-volume formulation. Second, finite-volume methods do not require a coordinate transformation in order to be applied on irregular meshes. As a result, they can be applied on *unstructured* meshes consisting of arbitrary polyhedra in three dimensions or arbitrary polygons in two dimensions. This increased flexibility can be advantageous in generating grids about complex geometries.

The PDE or divergence form of a conservation law can be written as

$$\frac{\partial Q}{\partial t} + \nabla \cdot \mathbf{F} = P, \tag{2.52}$$

where Q is a vector containing the set of variables which are conserved, e.g. mass, momentum, and energy, per unit volume, $\mathbf{F}$ is a set of vectors, or tensor, containing the flux of Q per unit area per unit time, P is the rate of production of Q per unit volume per unit time, and $\nabla \cdot \mathbf{F}$ is the well-known divergence operator. The same conservation law can be expressed in integral form as

$$\frac{\mathrm{d}}{\mathrm{d}t}\int_{V(t)} Q\mathrm{d}V + \oint_{S(t)} \mathbf{n} \cdot \mathbf{F}\mathrm{d}S = \int_{V(t)} P\mathrm{d}V. \tag{2.53}$$

This equation is a statement of the conservation of the conserved quantities in a finite region of space with volume $V(t)$ and surface area $S(t)$. In two dimensions, the region of space, or cell, is an area $A(t)$ bounded by a closed contour $C(t)$. The vector $\mathbf{n}$ is a unit vector normal to the surface pointing outward.

The basic idea of a finite-volume method is to satisfy the integral form of the conservation law to some degree of approximation for each of many contiguous control volumes that cover the domain of interest. Hence the function of the grid is to tessellate the domain into contiguous control volumes, and the volume V in (2.53) is that of a control volume whose shape is dependent on the nature of the grid. Examining (2.53), we see that several approximations must be made. The flux is required at the boundary of the control volume, which is a closed surface in three dimensions and a closed contour in two dimensions. This flux must then be integrated to find the net flux through the boundary. Similarly, the source term P

must be integrated over the control volume. Next a time-marching method can be applied to find the value of

$$\int_V Q \mathrm{d}V \tag{2.54}$$

at the next time step.

Let us consider each of these approximations in more detail. First, we note that the average value of Q in a cell with volume V is

$$\bar{Q} \equiv \frac{1}{V} \int_V Q \mathrm{d}V, \tag{2.55}$$

and (2.53) can be written as

$$V \frac{\mathrm{d}}{\mathrm{d}t} \bar{Q} + \oint_S \mathbf{n} \cdot \mathbf{F} \mathrm{d}S = \int_V P \mathrm{d}V \tag{2.56}$$

for a control volume that does not vary with time. Thus after applying a time-marching method, we have updated values of the cell-averaged quantities $\bar{Q}$. In order to evaluate the fluxes, which are a function of Q, at the control-volume boundary, Q can be represented within the cell by some piecewise approximation which produces the correct value of $\bar{Q}$. This is a form of interpolation often referred to as *reconstruction*. Each cell will have a different piecewise approximation to Q. When these are used to calculate $\mathbf{F}(Q)$, they will generally produce different approximations to the flux at the boundary between two control volumes, that is, the flux will be discontinuous. A nondissipative scheme analogous to centered differencing is obtained by taking the average of these two fluxes. Another approach known as flux-difference splitting is described in Sect. 2.5.

The basic elements of a finite-volume method are thus the following:

(1) Given the value of $\bar{Q}$ for each control volume, construct an approximation to $Q(x, y, z)$ in each control volume. Using this approximation, find Q at the control-volume boundary. Evaluate $\mathbf{F}(Q)$ at the boundary. Since there is a distinct approximation to $Q(x, y, z)$ in each control volume, two distinct values of the flux will generally be obtained at any point on the boundary between two control volumes.

(2) Apply some strategy for resolving the discontinuity in the flux at the control-volume boundary to produce a single value of $\mathbf{F}(Q)$ at any point on the boundary. This issue is discussed in Sect. 2.5.

(3) Integrate the flux to find the net flux through the control-volume boundary using some sort of quadrature.

(4) Advance the solution in time using a time-marching method to obtain new values of $\bar{Q}$.

The order of accuracy of the method is dependent on each of the approximations.

In order to include diffusive fluxes, the following relation between ∇Q and Q is sometimes used:

$$\int_V \nabla Q \mathrm{d}V = \oint_S \mathbf{n} Q \mathrm{d}S \tag{2.57}$$

or, in two dimensions,

$$\int_A \nabla Q \mathrm{d}A = \oint_C \mathbf{n} Q \mathrm{d}l, \tag{2.58}$$

where the unit vector **n** points outward from the surface or contour.

2.4.2 One-Dimensional Examples

We restrict our attention to a scalar dependent variable u and a scalar flux f, as in the model equations. We consider an equispaced grid with spacing Δx. The nodes of the grid are located at $x_j = j\Delta x$ as usual. Control volume j extends from $x_j - \Delta x/2$ to $x_j + \Delta x/2$, as shown in Fig. 2.4. This is referred to as a node centered scheme in contrast to a cell-centered scheme, where the control volume would extend from x_j to x_{j+1}. With respect to the discussion in this section, these two approaches are identical. We will use the following notation:

$$x_{j-1/2} = x_j - \Delta x/2, \quad x_{j+1/2} = x_j + \Delta x/2, \tag{2.59}$$

$$u_{j\pm1/2} = u(x_{j\pm1/2}), \quad f_{j\pm1/2} = f(u_{j\pm1/2}). \tag{2.60}$$

With these definitions, the cell-average value becomes

$$\bar{u}_j(t) \equiv \frac{1}{\Delta x}\int_{x_{j-1/2}}^{x_{j+1/2}} u(x, t)\mathrm{d}x, \tag{2.61}$$

and the integral form becomes

$$\frac{\mathrm{d}}{\mathrm{d}t}(\Delta x \bar{u}_j) + f_{j+1/2} - f_{j-1/2} = \int_{x_{j-1/2}}^{x_{j+1/2}} P\mathrm{d}x. \tag{2.62}$$

The integral form of the linear convection equation is obtained with $f = au$ and $P = 0$, while the integral form of the diffusion equation is obtained with $f = -\nu\nabla u = -\nu\partial u/\partial x$ and $P = 0$.

A Second-Order Approximation to the Convection Equation. With $a = 1$, the integral form of the linear convection equation becomes

$$\Delta x\frac{\mathrm{d}\bar{u}_j}{\mathrm{d}t} + f_{j+1/2} - f_{j-1/2} = 0 \tag{2.63}$$

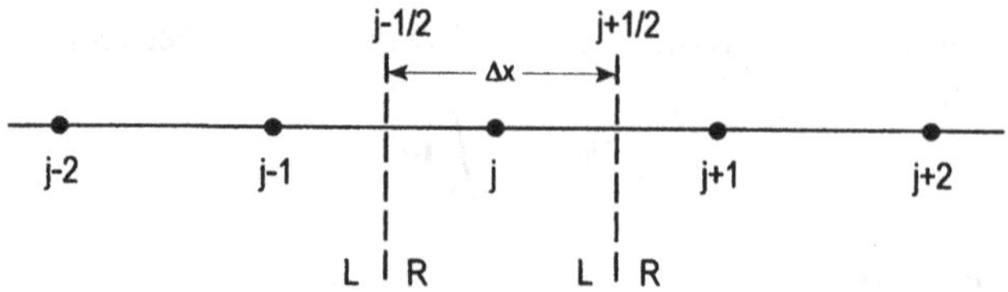

Fig. 2.4 Control volume in one dimension

with $f = u$. We choose a piecewise-constant approximation to $u(x)$ in each cell such that

$$u(x) = \bar{u}_j \quad x_{j-1/2} \leq x \leq x_{j+1/2}. \tag{2.64}$$

Evaluating this at $j + 1/2$ gives

$$f^{\mathrm{L}}_{j+1/2} = f(u^{\mathrm{L}}_{j+1/2}) = u^{\mathrm{L}}_{j+1/2} = \bar{u}_j, \tag{2.65}$$

where the L indicates that this approximation to $f_{j+1/2}$ is obtained from the approximation to $u(x)$ in the cell to the *left* of $x_{j+1/2}$, as shown in Fig. 2.4. The cell to the *right* of $x_{j+1/2}$, which is cell $j + 1$, gives

$$f^{\mathrm{R}}_{j+1/2} = \bar{u}_{j+1}. \tag{2.66}$$

Similarly, cell j is the cell to the right of $x_{j-1/2}$, giving

$$f^{\mathrm{R}}_{j-1/2} = \bar{u}_j \tag{2.67}$$

and cell $j - 1$ is the cell to the left of $x_{j-1/2}$, giving

$$f^{\mathrm{L}}_{j-1/2} = \bar{u}_{j-1}. \tag{2.68}$$

We have now accomplished the first step from the list in Sect. 2.4.1; we have defined the fluxes at the cell boundaries in terms of the cell-average data. In this example, the discontinuity in the flux at the cell boundary is resolved by taking the average of the fluxes on either side of the boundary. Thus

$$\hat{f}_{j+1/2} = \frac{1}{2}(f^{\mathrm{L}}_{j+1/2} + f^{\mathrm{R}}_{j+1/2}) = \frac{1}{2}(\bar{u}_j + \bar{u}_{j+1}) \tag{2.69}$$

and

$$\hat{f}_{j-1/2} = \frac{1}{2}(f^{\mathrm{L}}_{j-1/2} + f^{\mathrm{R}}_{j-1/2}) = \frac{1}{2}(\bar{u}_{j-1} + \bar{u}_j), \tag{2.70}$$

where $\hat{f}$ denotes a *numerical* flux which is an approximation to the exact flux.

Substituting (2.69) and (2.70) into the integral form, (2.63), we obtain

$$\begin{aligned}
&\Delta x \frac{\mathrm{d}\bar{u}_j}{\mathrm{d}t} + \frac{1}{2}(\bar{u}_j + \bar{u}_{j+1}) - \frac{1}{2}(\bar{u}_{j-1} + \bar{u}_j) \\
&= \Delta x \frac{\mathrm{d}\bar{u}_j}{\mathrm{d}t} + \frac{1}{2}(\bar{u}_{j+1} - \bar{u}_{j-1}) = 0.
\end{aligned} \tag{2.71}$$

With periodic boundary conditions, this point operator produces the following semi-discrete form:

$$\frac{\mathrm{d}\vec{\bar{u}}}{\mathrm{d}t} = -\frac{1}{2\Delta x} B_{\mathrm{p}}(-1, 0, 1)\vec{\bar{u}} \tag{2.72}$$

This is identical to the expression obtained using second-order centered differences, except it is written in terms of the cell average $\vec{\bar{u}}$, rather than the nodal values, $\vec{u}$. Hence our analysis and understanding of the eigensystem of the matrix $B_{\mathrm{p}}(-1, 0, 1)$ is relevant to finite-volume methods as well as finite-difference methods. Since the eigenvalues of $B_{\mathrm{p}}(-1, 0, 1)$ are pure imaginary, we can conclude that the use of the average of the fluxes on either side of the cell boundary, as in (2.69) and (2.70), leads to a nondissipative finite-volume method.

A Fourth-Order Approximation to the Convection Equation. A fourth-order spatial discretization can be obtained by replacing the piecewise-constant approximation in Sect. 2.4.2 with a piecewise-quadratic approximation as follows

$$u(\xi) = a\xi^2 + b\xi + c, \tag{2.73}$$

where ξ is again equal to $x - x_j$. The three parameters a, b, and c are chosen to satisfy the following constraints:

$$\begin{aligned}
\frac{1}{\Delta x}\int_{-3\Delta x/2}^{-\Delta x/2} u(\xi)\mathrm{d}\xi &= \bar{u}_{j-1} \\
\frac{1}{\Delta x}\int_{-\Delta x/2}^{\Delta x/2} u(\xi)\mathrm{d}\xi &= \bar{u}_j \\
\frac{1}{\Delta x}\int_{\Delta x/2}^{3\Delta x/2} u(\xi)\mathrm{d}\xi &= \bar{u}_{j+1}.
\end{aligned} \tag{2.74}$$

These constraints lead to

$$\begin{aligned}
a &= \frac{\bar{u}_{j+1} - 2\bar{u}_j + \bar{u}_{j-1}}{2\Delta x^2} \\
b &= \frac{\bar{u}_{j+1} - \bar{u}_{j-1}}{2\Delta x}
\end{aligned}$$

$$c = \frac{-\bar{u}_{j-1} + 26\bar{u}_j - \bar{u}_{j+1}}{24}. \tag{2.75}$$

It is left as an exercise for the reader to show that this reconstruction leads to the following form:

$$\Delta x \frac{\mathrm{d}\bar{u}_j}{\mathrm{d}t} + \frac{1}{12}(-\bar{u}_{j+2} + 8\bar{u}_{j+1} - 8\bar{u}_{j-1} + \bar{u}_{j-2}) = 0, \tag{2.76}$$

which is analogous to a fourth-order centered finite-difference scheme.
A Second-Order Approximation to the Diffusion Equation. In this section, we describe two approaches to deriving a finite-volume approximation to the diffusion equation. The first approach is simpler to extend to multidimensions, while the second approach is more suited to extension to higher-order accuracy.

With $\nu = 1$, the integral form of the diffusion equation is

$$\Delta x \frac{\mathrm{d}\bar{u}_j}{\mathrm{d}t} + f_{j+1/2} - f_{j-1/2} = 0 \tag{2.77}$$

with $f = -\nabla u = -\partial u/\partial x$. Also, (2.58) becomes

$$\int_a^b \frac{\partial u}{\partial x} \mathrm{d}x = u(b) - u(a). \tag{2.78}$$

We can thus write the following expression for the average value of the gradient of u over the interval $x_j \leq x \leq x_{j+1}$:

$$\frac{1}{\Delta x} \int_{x_j}^{x_{j+1}} \frac{\partial u}{\partial x} \mathrm{d}x = \frac{1}{\Delta x}(u_{j+1} - u_j). \tag{2.79}$$

The value of a continuous function at the center of a given interval is equal to the average value of the function over the interval to second-order accuracy. Hence, to second-order, we can write

$$\hat{f}_{j+1/2} = -\left(\frac{\partial u}{\partial x}\right)_{j+1/2} = -\frac{1}{\Delta x}(\bar{u}_{j+1} - \bar{u}_j). \tag{2.80}$$

Similarly,

$$\hat{f}_{j-1/2} = -\frac{1}{\Delta x}(\bar{u}_j - \bar{u}_{j-1}). \tag{2.81}$$

Substituting these into the integral form (2.77), we obtain

$$\Delta x \frac{\mathrm{d}\bar{u}_j}{\mathrm{d}t} = \frac{1}{\Delta x}(\bar{u}_{j-1} - 2\bar{u}_j + \bar{u}_{j+1}) \tag{2.82}$$

or, with Dirichlet boundary conditions,

$$\frac{\mathrm{d}\vec{\bar{u}}}{\mathrm{d}t} = \frac{1}{\Delta x^2} B(1, -2, 1)\vec{\bar{u}} + \left(\vec{bc}\right). \tag{2.83}$$

This provides a semi-discrete finite-volume approximation to the diffusion equation, and we see that the properties of the matrix $B(1, -2, 1)$ are relevant to the study of finite-volume methods as well as finite-difference methods.

For our second approach, we use a piecewise-quadratic approximation as in Sect. 2.4.2. From (2.73) we have

$$\frac{\partial u}{\partial x} = \frac{\partial u}{\partial \xi} = 2a\xi + b \tag{2.84}$$

with a and b given in (2.75). With $f = -\partial u/\partial x$, this gives

$$f^{\mathrm{R}}_{j+1/2} = f^{\mathrm{L}}_{j+1/2} = -\frac{1}{\Delta x}(\bar{u}_{j+1} - \bar{u}_j) \tag{2.85}$$

$$f^{\mathrm{R}}_{j-1/2} = f^{\mathrm{L}}_{j-1/2} = -\frac{1}{\Delta x}(\bar{u}_j - \bar{u}_{j-1}). \tag{2.86}$$

Notice that there is no discontinuity in the flux at the cell boundary. This produces

$$\frac{\mathrm{d}\bar{u}_j}{\mathrm{d}t} = \frac{1}{\Delta x^2}(\bar{u}_{j-1} - 2\bar{u}_j + \bar{u}_{j+1}), \tag{2.87}$$

which is identical to (2.82).

2.5 Numerical Dissipation and Upwind Schemes

For a given order of accuracy, centered difference schemes produce the lowest coefficient of the leading truncation error term in comparison with one-sided and biased schemes. Moreover, a centered difference approximation correctly mimics the physics of convection and diffusion. In particular, a centered approximation to a first derivative is nondissipative, i.e. the eigenvalues of the associated matrix operator are pure imaginary. No aphysical numerical dissipation is introduced. Nevertheless, in the numerical solution of many practical problems, a small well-controlled amount of numerical dissipation is desirable and possibly even necessary for stability.

In a linear problem, there exist modes that are inaccurately resolved, as demonstrated by the modified wavenumbers shown in Fig. 2.2. If these modes are introduced into a simulation somehow, for example by the initial conditions, and there exists no mechanism to damp them, then they will persist and potentially contaminate the solution. It is preferable to damp these under-resolved solution components. In processes governed by nonlinear equations, such as the Navier–Stokes equations, there can be a continual production of high-frequency components of the solution, leading, for example, to the formation of shock waves. In a real physical problem, the production of high frequencies is eventually limited by viscosity. However, in practical simulations, the smallest length scales where the physical damping occurs are often under resolved. Unless the relevant length scales are resolved, some form of added numerical dissipation is required. Since the addition of numerical dissipation is tantamoun to intentionally introducing nonphysical behavior, it must be carefully controlled such that the error introduced is not excessive.

2.5.1 Numerical Dissipation in the Linear Convection Equation

One means of introducing numerical dissipation is through the use of one-sided differencing in the inviscid flux terms. For example, consider the following point operator for the spatial derivative term in the linear convection equation:

$$\begin{aligned} -a(\delta_x u)_j &= \frac{-a}{2\Delta x}[-(1+\beta)u_{j-1} + 2\beta u_j + (1-\beta)u_{j+1}] \\ &= \frac{-a}{2\Delta x}[(-u_{j-1} + u_{j+1}) + \beta(-u_{j-1} + 2u_j - u_{j+1})]. \end{aligned} \tag{2.88}$$

The second form shown divides the operator into an antisymmetric component $(-u_{j-1} + u_{j+1})/2\Delta x$ and a symmetric component $\beta(-u_{j-1} + 2u_j - u_{j+1})/2\Delta x$. The antisymmetric component is the second-order centered difference operator. With $\beta \neq 0$, the operator is only first-order accurate. A *backward* difference operator is given by $\beta = 1$, and a *forward* difference operator is given by $\beta = -1$.

For periodic boundary conditions, the corresponding matrix operator is

$$-a\delta_x = \frac{-a}{2\Delta x}B_p(-1-\beta, 2\beta, 1-\beta).$$

The eigenvalues of this matrix are

$$\lambda_m = \frac{-a}{\Delta x}\left\{\beta\left[1 - \cos\left(\frac{2\pi m}{M}\right)\right] + \mathrm{i}\sin\left(\frac{2\pi m}{M}\right)\right\}, \quad m = 0, 1, \ldots, M-1.$$

If a is positive, the forward difference operator ($\beta = -1$) produces $\Re(\lambda_m) > 0$, the centered difference operator ($\beta = 0$) produces $\Re(\lambda_m) = 0$, and the backward difference operator produces $\Re(\lambda_m) < 0$. Hence the forward difference operator is inherently unstable, while the centered and backward operators are inherently stable. If a is negative, the roles are reversed.

In order to devise a spatial discretization that is stable independent of the sign of a, we can rewrite the linear convection equation as

$$\frac{\partial u}{\partial t} + (a^+ + a^-)\frac{\partial u}{\partial x} = 0, \quad a^\pm = \frac{a \pm |a|}{2}. \tag{2.89}$$

If $a \geq 0$, then $a^+ = a \geq 0$ and $a^- = 0$. Alternatively, if $a \leq 0$, then $a^+ = 0$ and $a^- = a \leq 0$. Now we can safely use a backward difference approximation for the a^+ (≥ 0) term and a forward difference approximation for the a^- (≤ 0) term. This is the basic concept behind upwind methods, that is some decomposition or splitting of the fluxes into terms which have positive and negative characteristic speeds so that appropriate differencing schemes can be chosen for each.

The above approach can be written in a different, but entirely equivalent, manner. From (2.88), we see that a stable discretization is obtained with $\beta = 1$ if $a \geq 0$ and with $\beta = -1$ if $a \leq 0$. This is achieved by the following point operator:

$$-a(\delta_x u)_j = \frac{-1}{2\Delta x}[a(-u_{j-1} + u_{j+1}) + |a|(-u_{j-1} + 2u_j - u_{j+1})]. \tag{2.90}$$

Any symmetric component in the spatial operator introduces dissipation (or amplification). Therefore, one could choose $\beta = 1/2$ in (2.88), for example, leading to the following operator:

$$-a(\delta_x u)_j = \frac{-1}{2\Delta x}[a(-u_{j-1} + u_{j+1}) + \frac{1}{2}|a|(-u_{j-1} + 2u_j - u_{j+1})]. \tag{2.91}$$

The resulting spatial operator is not one-sided, but it is dissipative.

Similarly, biased schemes use more information on one side of the grid node than the other. For example, a third-order backward-biased scheme is given by

$$\begin{aligned}(\delta_x u)_j &= \frac{1}{6\Delta x}(u_{j-2} - 6u_{j-1} + 3u_j + 2u_{j+1}) \\ &= \frac{1}{12\Delta x}[(u_{j-2} - 8u_{j-1} + 8u_{j+1} - u_{j+2}) \\ &\quad + (u_{j-2} - 4u_{j-1} + 6u_j - 4u_{j+1} + u_{j+2})].\end{aligned} \tag{2.92}$$

The antisymmetric component of this operator is the fourth-order centered difference operator. The symmetric component approximates $\Delta x^3 u_{xxxx}/12$. Therefore, this operator produces fourth-order accuracy in phase, with a third-order dissipative term. Note that the antisymmetric portion of the first-derivative operator always has an even order of accuracy, while the symmetric portion always has an odd order.

2.5.2 Upwind Schemes

In Sect. 2.5.1, we saw that numerical dissipation can be introduced in the spatial difference operator by using one-sided difference schemes or, more generally, by adding a symmetric component to the spatial operator. With this approach, the direction of the one-sided operator (i.e. whether it is a forward or a backward difference) and the sign of the symmetric component depend on the sign of the wave speed. When a *hyperbolic system* of equations is being solved, the wave speeds can be both positive and negative. For example, the eigenvalues of the flux Jacobian for the one-dimensional Euler equations are $u, u+a, u-a$, where u is the fluid velocity and a is the speed of sound. When the flow is subsonic, these are of mixed sign. In order to apply one-sided differencing schemes to such systems, some form of splitting is required.

Flux-Vector Splitting. Consider a linear, constant-coefficient, hyperbolic system of partial differential equations given by

$$\frac{\partial u}{\partial t}+\frac{\partial f}{\partial x}=\frac{\partial u}{\partial t}+A\frac{\partial u}{\partial x}=0, \tag{2.93}$$

where $f = Au$, and A is diagonalizable with real eigenvalues. This system can be decoupled into characteristic equations of the form

$$\frac{\partial w_i}{\partial t}+\lambda_i\frac{\partial w_i}{\partial x}=0, \tag{2.94}$$

where the wave speeds, λ_i, are the eigenvalues of the Jacobian matrix, A, and the w_is are the characteristic variables. In order to apply a one-sided (or biased) spatial differencing scheme, we need to apply a backward difference if the wave speed, λ_i, is positive, and a forward difference if the wave speed is negative.

To accomplish this, we split the matrix of eigenvalues, Λ, into two components such that

$$\Lambda=\Lambda^{+}+\Lambda^{-}, \tag{2.95}$$

where

$$\Lambda^{+}=\frac{\Lambda+|\Lambda|}{2},\quad \Lambda^{-}=\frac{\Lambda-|\Lambda|}{2}. \tag{2.96}$$

With these definitions, Λ^{+} contains the positive eigenvalues and Λ^{-} contains the negative eigenvalues. With the additional definitions[2]

$$A^{+}=X\Lambda^{+}X^{-1},\quad A^{-}=X\Lambda^{-}X^{-1}, \tag{2.97}$$

we can define the split flux vectors as

[2] With these definitions, A^{+} has all nonnegative eigenvalues, and A^{-} has all nonpositive eigenvalues.

$$f^+ = A^+ u, \quad f^- = A^- u. \tag{2.98}$$

Noting that $f = f^+ + f^-$, we can rewrite the original system in terms of the split flux vectors as

$$\frac{\partial u}{\partial t} + \frac{\partial f^+}{\partial x} + \frac{\partial f^-}{\partial x} = 0. \tag{2.99}$$

The spatial terms have been split into two components according to the sign of the wave speeds. A dissipative scheme is obtained by applying backward differencing to the $\frac{\partial f^+}{\partial x}$ term and forward differencing to the $\frac{\partial f^-}{\partial x}$ term.

Flux-vector splitting [2, 3] can also be used with a finite-volume method. Referring back to Sect. 2.4, recall that in a finite-volume method there exists a discontinuity in the flux at a control-volume boundary. When we took the average of the two fluxes at the interface, we obtained a nondissipative finite-volume discretization analogous to centered differencing. In order to develop a dissipative scheme, we can instead choose f^+ from the state to the left of the interface and f^- from the right state. This leads to the following upwind numerical flux:

$$\hat{f}_{j+1/2} = (f^+)^L + (f^-)^R, \tag{2.100}$$

which leads to a finite-volume method that is analogous to the flux-vector-split finite-difference scheme described above.

Flux-Difference Splitting. With flux-difference splitting [4], the numerical flux is given by

$$\hat{f}_{j+1/2} = \frac{1}{2}\left(f^L + f^R\right) + \frac{1}{2}|A|\left(u^L - u^R\right), \tag{2.101}$$

where

$$|A| = X|\Lambda|X^{-1}. \tag{2.102}$$

It is straightforward to show that in the linear, constant-coefficient case this is entirely equivalent to (2.100).

2.5.3 Artificial Dissipation

We have seen that numerical dissipation can be introduced by using one-sided differencing schemes together with some form of flux splitting. We have also seen that such dissipation can be introduced by adding a symmetric component to an antisymmetric (dissipation-free) operator. Thus we can generalize the concept of upwinding to include any scheme in which the symmetric portion of the operator is treated in such a manner as to be truly dissipative.

For example, consider the operator

$$\delta_x f = \delta_x^{\mathrm{a}} f + \delta_x^{\mathrm{s}}(|A|u), \tag{2.103}$$

where δ_x^{a} and δ_x^{s} are antisymmetric and symmetric difference operators, and $|A|$ is defined in (2.102). The second spatial term is known as *artificial dissipation*. With appropriate choices of δ_x^{a} and δ_x^{s}, this approach can be identical to the upwind approach.

It is common to use the following operator for δ_x^{s}

$$\left(\delta_x^{\mathrm{s}} u\right)_j = \frac{\epsilon}{\Delta x}(u_{j-2} - 4u_{j-1} + 6u_j - 4u_{j+1} + u_{j+2}), \tag{2.104}$$

where ϵ is a problem-dependent coefficient. This symmetric operator approximates $\epsilon \Delta x^3 u_{xxxx}$ and thus introduces a third-order dissipative term. With an appropriate value of ϵ, this often provides sufficient damping of high frequency modes without greatly affecting the low frequency modes. A more complicated treatment of the numerical dissipation is required near shock waves and other discontinuities; this subject is dealt with in later chapters.

2.6 Time-Marching Methods for ODEs

2.6.1 Basic Concepts: Explicit and Implicit Methods

After discretizing the spatial derivatives in the governing PDEs (such as the Navier–Stokes equations), we obtain a coupled system of nonlinear ODEs in the form

$$\frac{\mathrm{d}\vec{u}}{\mathrm{d}t} = \vec{F}(\vec{u}, t). \tag{2.105}$$

These can be integrated in time using a time-marching method to obtain a time-accurate solution to an *unsteady* flow problem. For a *steady* flow problem, spatial discretization leads to a coupled system of nonlinear algebraic equations in the form

$$\vec{F}(\vec{u}) = 0. \tag{2.106}$$

As a result of the nonlinearity of these equations, some sort of iterative method is required to obtain a solution. For example, one can consider the use of Newton's method, which is widely used for nonlinear algebraic equations. This produces an iterative method in which a coupled system of linear algebraic equations must be solved at each iteration. Alternatively, one can consider a time-dependent path to the steady state and use a time-marching method to integrate the unsteady form of the equations until the solution is sufficiently close to the steady solution. The subject of the present section, time-marching methods for ODEs, is thus relevant to both steady and unsteady flow problems. When using a time-marching method to compute steady flows, the goal is simply to remove the transient portion of the solution as quickly as possible; time-accuracy is not required. This motivates the study of stability and stiffness, topics which are discussed in the next section.

Application of a time-marching method to an ODE produces an ordinary *difference* equation (OΔE). Simple OΔEs can be easily solved, so we can develop exact solutions for the model OΔEs arising from the application of time-marching methods to the model ODEs. Using these exact solutions, we can analyze and understand the stability and accuracy properties of various time-marching methods.

Based on the discussion in Sect. 2.3, we will consider scalar ODEs given by

$$\frac{\mathrm{d}u}{\mathrm{d}t} = u' = F(u, t), \tag{2.107}$$

bearing in mind that the analysis applies directly to the solution of systems of ODEs. As in Sect. 2.2, we use the convention that the n subscript, or the (n) superscript, always denotes a discrete time value, and h represents the time interval Δt. Combining this notation with (2.107) gives

$$u'_n = F_n = F(u_n, t_n), \quad t_n = nh.$$

Often we need a more sophisticated notation for intermediate time steps involving intermediate solutions denoted by $\tilde{u}$, $\bar{u}$, etc. For these we use the notation

$$\tilde{u}'_{n+\alpha} = \tilde{F}_{n+\alpha} = F(\tilde{u}_{n+\alpha}, t_n + \alpha h).$$

The methods we study are to be applied to linear or nonlinear ODEs, but the methods themselves are formed by linear combinations of the dependent variable and its derivative at various time intervals. They are represented conceptually by

$$u_{n+1} = f\left(\beta_1 h u'_{n+1}, \beta_0 h u'_n, \beta_{-1} u'_{n-1}, \ldots, \alpha_0 u_n, \alpha_{-1} u_{n-1}, \ldots\right). \tag{2.108}$$

With an appropriate choice of the αs and βs, these methods can be constructed to give a local Taylor series accuracy of any order. A method is said to be *explicit* if $\beta_1 = 0$ and *implicit* otherwise. An *explicit* method is one in which the new predicted solution is only a function of known data, for example, u'_n, u'_{n-1}, u_n, and u_{n-1} for a method using two previous time levels, and therefore the time advance is simple. For an *implicit* method, the new predicted solution is also a function of the time derivative at the new time level, that is, u'_{n+1}. As we shall see, for systems of ODEs and nonlinear problems, implicit methods require more complicated strategies to solve for u_{n+1} than explicit methods.

Most time-marching methods in current use in CFD fall into one of three categories: *linear multistep methods*, *predictor–corrector methods*, and *Runge–*

Kutta methods. For the purpose of our analysis, we group predictor–corrector and Runge–Kutta methods together under the heading *multi–stage methods.*

In a linear multistep method, the solution at the new time level is a linear combination of the solution and its derivative at various time levels. In other words, (2.108) becomes

$$u_{n+1} = \beta_1 hu'_{n+1} + \beta_0 hu'_n, \beta_{-1}u'_{n-1} + \cdots + \alpha_0 u_n + \alpha_{-1}u_{n-1} + \cdots \quad (2.109)$$

Specific linear multistep methods are associated with specific choices of the αs and βs. In order to establish the order of accuracy, one can perform a Taylor series expansion of the right-hand side of (2.109) with particular values of the αs and βs and compare it to the Taylor series expansion of u_{n+1}. The order of accuracy of the method is the lowest exponent of h in the difference minus one. Similarly, one can use a Taylor table to derive a method by choosing which αs and βs will be permitted to have nonzero values; the Taylor table facilitates the derivation of the α and β values that maximize the order of accuracy. For example, the most basic time-marching method, which we will call the explicit Euler method, is found with all αs and βs set to zero with the exception of α_0 and β_0. In order to maximize the order of accuracy, one must choose $\alpha_0 = \beta_0 = 1$, which gives

$$u_{n+1} = u_n + hu'_n + O(h^2). \quad (2.110)$$

Since the leading error term is $O(h^2)$, this method is first order. This means that if an ODE is solved with this method over a specific time interval using first a specific value of h and then with $h/2$, the error in the solution at the end of the time interval will be reduced by a factor of two.

Further examples of linear multistep methods commonly used in CFD applications are given below[3]:

Explicit Methods.

$$\begin{aligned} u_{n+1} &= u_{n-1} + 2hu'_n && \text{Leapfrog} \\ u_{n+1} &= u_n + \tfrac{1}{2}h\big[3u'_n - u'_{n-1}\big] && \text{AB2} \\ u_{n+1} &= u_n + \tfrac{h}{12}\big[23u'_n - 16u'_{n-1} + 5u'_{n-2}\big] && \text{AB3} \end{aligned}$$

Implicit Methods.

$$\begin{aligned} u_{n+1} &= u_n + hu'_{n+1} && \text{Implicit Euler} \\ u_{n+1} &= u_n + \tfrac{1}{2}h\big[u'_n + u'_{n+1}\big] && \text{Trapezoidal (AM2)} \\ u_{n+1} &= \tfrac{1}{3}\big[4u_n - u_{n-1} + 2hu'_{n+1}\big] && \text{2nd-order Backward} \\ u_{n+1} &= u_n + \tfrac{h}{12}\big[5u'_{n+1} + 8u'_n - u'_{n-1}\big] && \text{AM3} \end{aligned}$$

[3] Where the notation AB2 refers to the 2nd-order Adams-Bashforth method and AM2 refers to the second-order Adams-Moulton method, etc.

Predictor–corrector methods constructed to time-march linear or nonlinear ODEs are composed of sequences of linear multistep methods, each of which is referred to as a family in the solution process. There may be many families in the sequence, and usually the final family has a higher Taylor-series order of accuracy than the intermediate ones. Their use is motivated by ease of application and increased efficiency. In a simple two–stage predictor–corrector method, the solution is initially *predicted* at the next time level or some intermediate time using a linear multistep method. It is then *corrected* by applying another linear multistep method that involves applying the derivative function $F(u, t)$ at the predicted u and the appropriate value of t. For example, the second-order predictor–corrector method we will call MacCormack's method[4] can be written as

$$\begin{aligned} \tilde{u}_{n+1} &= u_n + hu'_n \\ u_{n+1} &= \frac{1}{2}(u_n + \tilde{u}_{n+1} + h\tilde{u}'_{n+1}). \end{aligned} \tag{2.111}$$

The predicted solution $\tilde{u}_{n+1}$ is obtained at t_{n+1} using the explicit Euler method, while the correction is obtained using the implicit trapezoidal method (see examples above) with u'_{n+1} replaced by $\tilde{u}'_{n+1}$. The method is explicit, since $\tilde{u}_{n+1}$ is computed before $\tilde{u}'_{n+1}$ is needed. Note that in order to advance one time step, two evaluations of the derivative function $F(u, t)$ are required, $F(u_n, t_n)$ in the predictor and $F(\tilde{u}_{n+1}, t_{n+1})$ in the corrector. Since evaluating the derivative function is typically the greatest computing expense in the application of a time-marching method, this means that the cost per time step of MacCormack's method is nominally twice that of a linear multistep method, where only one derivative function evaluation is needed per time step.[5]

Runge–Kutta methods are another important subset of multi-stage methods. The most popular is the classical explicit fourth-order Runge–Kutta method, which can be written in a notation consistent with the predictor–corrector example as

$$\begin{aligned} \widehat{u}_{n+1/2} &= u_n + \frac{1}{2}hu'_n \\ \tilde{u}_{n+1/2} &= u_n + \frac{1}{2}h\widehat{u}'_{n+1/2} \\ \overline{u}_{n+1} &= u_n + h\tilde{u}'_{n+1/2} \\ u_{n+1} &= u_n + \frac{1}{6}h\left[u'_n + 2\left(\widehat{u}'_{n+1/2} + \tilde{u}'_{n+1/2}\right) + \overline{u}'_{n+1}\right]. \end{aligned} \tag{2.112}$$

[4] Here we discuss only MacCormack's *time-marching* method. The method commonly referred to as MacCormack's method is a fully-discrete method [5].

[5] If a linear multistep method requires, for example, $F(u_{n-1}, t_{n-1})$, this can be calculated at a previous time step and stored.

This method requires four derivative function evaluations per time step. As described below, the analysis and derivation of multi-stage methods is more involved than that for linear multistep methods.

2.6.2 Converting Time-Marching Methods to OΔEs

In Sect. 2.3.5 we chose a representative scalar ODE for the study of time-marching methods given by

$$\frac{\mathrm{d}u}{\mathrm{d}t} = \lambda u + a\mathrm{e}^{\mu t}, \tag{2.113}$$

where λ, a, and μ are complex constants. This equation has the following exact solution (for $\mu \neq \lambda$):

$$u(t) = c\,\mathrm{e}^{\lambda t} + \frac{a\mathrm{e}^{\mu t}}{\mu - \lambda}, \tag{2.114}$$

where the constant c is determined from the initial condition. Of course, one would not normally apply a numerical method to solve an equation for which one can derive the exact solution. Our purpose here is to analyze and evaluate time-marching methods, and a model ODE with a known solution plays an important role in this process. Using the theory of OΔEs we can obtain a closed form solution for the *numerical* solution obtained when a given time-marching method is used to solve the representative ODE. Rather than having to conduct a series of numerical experiments in order to understand the properties of a time-marching method, we can use this closed form solution to obtain these properties as an explicit function of the parameters h, λ, a, and μ. Hence, the theory of OΔEs provides a powerful tool for analyzing and deriving time-marching methods.

For example, consider the application of the explicit Euler method (2.110) to the representative ODE. Noting that $t_n = hn$, one obtains

$$\begin{aligned} u_{n+1} &= u_n + h(\lambda u_n + a\mathrm{e}^{\mu hn}) \\ &= (1 + \lambda h)u_n + ha\mathrm{e}^{\mu hn}. \end{aligned} \tag{2.115}$$

This is a first-order inhomogenous OΔE that can be written in the general form

$$u_{n+1} = \sigma u_n + \hat{a}b^n, \tag{2.116}$$

where σ, $\hat{a}$, and b are, in general, complex parameters. The independent variable is n rather than t, and, since the equations are linear and have constant coefficients, σ is not a function of either n or u. The exact solution of (2.116) is (for $b \neq \sigma$):

$$u_n = c_1\sigma^n + \frac{\hat{a}b^n}{b-\sigma}, \tag{2.117}$$

where c_1 is a constant determined by the initial conditions. That (2.117) is a solution to (2.116) can be easily verified by substitution, and the reader is encouraged to do so.

Applying the exact OΔE solution (2.117) to the OΔE obtained by applying the explicit Euler method to the representative ODE (2.115), one obtains the exact numerical solution:

$$u_n = c_1(1+\lambda h)^n + \frac{hae^{\mu hn}}{e^{\mu h} - 1 - \lambda h}. \tag{2.118}$$

This can be compared directly with the exact ODE solution rewritten as

$$u(t) = c\,(e^{\lambda h})^n + \frac{ae^{\mu hn}}{\mu - \lambda}. \tag{2.119}$$

In particular, comparing the homogeneous solutions

$$c_1(1+\lambda h)^n \approx c\,(e^{\lambda h})^n, \tag{2.120}$$

where $c_1 = c$, shows that $\sigma = 1 + \lambda h$ is an approximation to $e^{\lambda h}$. Given that the Taylor series expansion of $e^{\lambda h}$ about $\lambda h = 0$ is

$$e^{\lambda h} = 1 + \lambda h + \frac{1}{2}\lambda^2 h^2 + \cdots + \frac{1}{k!}\lambda^k h^k + \cdots , \tag{2.121}$$

the error in the approximation is $O(h^2)$, consistent with the fact that the explicit Euler method is a first-order method. With a little more algebra,[6] one can readily show that the particular solution in (2.118) is also a first-order approximation to the exact particular solution.

Let us examine the homogeneous OΔE solution in more detail. Consider as an example $\lambda = -1$. The exact ODE homogeneous solution is simply ce^{-t}. The homogeneous solution for the explicit Euler OΔE is

$$u_n = c_1(1-h)^n. \tag{2.122}$$

For small h this is a good approximation, consistent with the fact that $\sigma \approx e^{\lambda h}$. However, for $h = 1$, the homogeneous solution becomes $u_n = 0$ after one step. This is completely inaccurate but at least provides the correct homogeneous solution as $n \to \infty$. With $h = 2$, the solution oscillates between 1 and -1, and for $h > 2$,

[6] One must expand both the exact ODE particular solution and the exact OΔE particular solution in Taylor series and compare on a term by term basis starting with the lowest power of h.

the solution grows without bound as $n \to \infty$. Generalizing this to arbitrary λ, the solution grows without bound if $|\sigma| = |1 + \lambda h| > 1$.[7]

Next consider the application of the implicit Euler method

$$u_{n+1} = u_n + hu'_{n+1} \tag{2.123}$$

to the representative ODE. The resulting OΔE is

$$u_{n+1} = \frac{1}{1-\lambda h}u_n + \frac{1}{1-\lambda h}he^{\mu h}ae^{\mu hn}. \tag{2.124}$$

This can once again be compared to the form (2.116) to obtain the exact OΔE solution

$$u_n = c_1\left(\frac{1}{1-\lambda h}\right)^n + ae^{\mu hn}\cdot\frac{he^{\mu h}}{(1-\lambda h)e^{\mu h}-1}. \tag{2.125}$$

In this case $\sigma = 1/(1-\lambda h)$. Although again a first-order approximation to $e^{\lambda h}$, this leads to quite different behaviour than $\sigma = 1 + \lambda h$ obtained for the explicit Euler method. For example, with $\lambda = -1$ as in our previous example, the solution will not become unbounded even as $h \to \infty$.

This approach based on (2.116) and its solution (2.117) enables us to study one-step linear multistep methods, which are linear multistep methods that use data only at time levels $n+1$ and n. For linear multistep methods of two steps or more and multistage methods, a more general theory is needed. This is achieved by writing the OΔE obtained by applying a time-marching method to the representative ODE in the following *operational form*:

$$P(E)u_n = Q(E)\cdot ae^{\mu hn}. \tag{2.126}$$

The terms $P(E)$ and $Q(E)$ are polynomials in E referred to as the *characteristic polynomial* and the *particular polynomial*, respectively. The *shift operator E* is defined formally by the relations

$$u_{n+1} = Eu_n, \quad u_{n+k} = E^k u_n$$

and also applies to exponents, thus

$$b^\alpha \cdot b^n = b^{n+\alpha} = E^\alpha \cdot b^n,$$

where α can be any fraction or irrational number.

[7] Recall that λ and hence σ are in general complex.

The general solution of (2.126) can be expressed as

$$u_n = \sum_{k=1}^{K} c_k(\sigma_k)^n + a\mathrm{e}^{\mu hn} \cdot \frac{Q(\mathrm{e}^{\mu h})}{P(\mathrm{e}^{\mu h})}, \tag{2.127}$$

where σ_k are the K roots of the characteristic polynomial, $P(\sigma) = 0$. An important subset of this solution occurs when $\mu = 0$, representing a time-invariant particular solution, or a steady state. In such a case

$$u_n = \sum_{k=1}^{K} c_k(\sigma_k)^n + a \cdot \frac{Q(1)}{P(1)}. \tag{2.128}$$

We shall illustrate the application of (2.126) and (2.127) with two examples, a two-step multistep method and a multistage method, MacCormack's predictor-corrector method (2.111).

Consider first the *leapfrog method*, a second-order explicit two-step multistep method given by[8]

$$u_{n+1} = u_{n-1} + 2hu'_n. \tag{2.129}$$

Applying it to the representative ODE gives

$$u_{n+1} = u_{n-1} + 2h(\lambda u_n + a\mathrm{e}^{\mu hn}). \tag{2.130}$$

After rearranging and introducing the shift operator ($u_{n+1} = Eu_n$, $u_{n-1} = E^{-1}u_n$,) we obtain

$$(E - 2\lambda h - E^{-1})u_n = 2ha\mathrm{e}^{\mu hn}, \tag{2.131}$$

which is in the form (2.126) with

$$P(E) = E - 2\lambda h - E^{-1}, \quad Q(E) = 2h. \tag{2.132}$$

Setting $P(\sigma) = 0$ gives the relation

$$\sigma^2 - 2\lambda h\sigma - 1 = 0, \tag{2.133}$$

which produces two σ roots:

$$\sigma_{1,2} = \lambda h \pm \sqrt{\lambda^2 h^2 + 1}. \tag{2.134}$$

[8] The reader should observe the relationship between this time-marching method and the second-order centered difference approximation to a first derivative.

Thus the OΔE solution is

$$u_n = c_1(\lambda h + \sqrt{\lambda^2 h^2 + 1})^n + c_2(\lambda h - \sqrt{\lambda^2 h^2 + 1})^n$$
$$+ae^{\mu hn} \cdot \frac{2h}{e^{\mu h} - 2\lambda h - e^{-\mu h}}. \tag{2.135}$$

This OΔE solution has an important difference from that obtained for the explicit and implicit Euler methods: two σ-roots. Only one of them approximates $e^{\lambda h}$. In this case $\sigma_1 = \lambda h + \sqrt{\lambda^2 h^2 + 1}$ can be expanded in a Taylor series to show that it is a second-order approximation to $e^{\lambda h}$. The root with this property is known as the *principal root*, and the other root or roots are known as *spurious roots*. There are two constants in the OΔE solution, but only one initial condition. This reflects the fact that a method requiring data at time level $n - 1$ or earlier is not self starting. At the first time step $n = 0$, $u_n = u_0$ is known from the initial condition, but u_{n-1} is not known. Therefore, such methods are normally started using a self-starting method for the first step or steps, as required, and this provides the second necessary constant. If the method is started in this manner, the coefficients of the spurious roots will have small (but not zero) magnitudes.

As our final example, we will derive the solution to the OΔE obtained by applying MacCormack's explicit predictor–corrector method (2.111) to the representative ODE. This methodology can be followed to analyze Runge-Kutta methods as well.[9] Applying MacCormack's method to the representative equation gives

$$\begin{aligned} \tilde{u}_{n+1} - (1 + \lambda h)u_n &= ahe^{\mu hn} \\ -\tfrac{1}{2}(1 + \lambda h)\tilde{u}_{n+1} + u_{n+1} - \tfrac{1}{2}u_n &= \tfrac{1}{2}ahe^{\mu h(n+1)}, \end{aligned} \tag{2.136}$$

which is a coupled set of linear OΔEs with constant coefficients. The second line in (2.136) is obtained by noting that

$$\begin{aligned} \tilde{u}'_{n+1} &= F(\tilde{u}_{n+1}, t_n + h) \\ &= \lambda \tilde{u}_{n+1} + ae^{\mu h(n+1)}. \end{aligned} \tag{2.137}$$

Introducing the shift operator E, we obtain

$$\begin{bmatrix} E & -(1 + (e^{\mu h})) \\ -\tfrac{1}{2}(1 + (e^{\mu h}))E & E - \tfrac{1}{2} \end{bmatrix} \begin{bmatrix} \tilde{u} \\ u \end{bmatrix}_n = h \cdot \begin{bmatrix} 1 \\ \tfrac{1}{2}E \end{bmatrix} \tilde{u}. \tag{2.138}$$

This system has a solution for both the intermediate family $\tilde{u}_n$ and the final family u_n. Since we are interested only in the final family, we can use Cramer's rule to obtain the operational form (2.126) as follows:

[9] In fact, MacCormack's method can be considered a second-order Runge-Kutta method.

$$P(E) = \det \begin{bmatrix} E & -(1+\lambda h) \\ -\frac{1}{2}(1+\lambda h)E & E - \frac{1}{2} \end{bmatrix} = E\left(E - 1 - \lambda h - \frac{1}{2}\lambda^2 h^2\right)$$

$$Q(E) = \det \begin{bmatrix} E & h \\ -\frac{1}{2}(1+\lambda h)E & \frac{1}{2}hE \end{bmatrix} = \frac{1}{2}hE(E + 1 + \lambda h).$$

The σ-root is found from

$$P(\sigma) = \sigma\left(\sigma - 1 - \lambda h - \frac{1}{2}\lambda^2 h^2\right) = 0,$$

which has only one nontrivial root

$$\sigma = 1 + \lambda h + \frac{1}{2}\lambda^2 h^2. \tag{2.139}$$

The complete solution can therefore be written

$$u_n = c_1\left(1 + \lambda h + \frac{1}{2}\lambda^2 h^2\right)^n + a\mathrm{e}^{\mu hn} \cdot \frac{\frac{1}{2}h\left(\mathrm{e}^{\mu h} + 1 + \lambda h\right)}{\mathrm{e}^{\mu h} - 1 - \lambda h - \frac{1}{2}\lambda^2 h^2}. \tag{2.140}$$

The σ-root is clearly a second-order approximation to $\mathrm{e}^{\lambda h}$, and the particular solution can also be shown to be a second-order approximation of the particular solution in (2.114). This example provides a template that can be used for the derivation and analysis of predictor–corrector and Runge-Kutta methods up to third order. Runge-Kutta methods of order four and higher must be derived based on a nonlinear ODE.

We are now in a position to generalize what we have learned about the σ-roots associated with a time-marching method. Recall that we intend to apply time-marching methods to systems of ODEs generated by discretizing the spatial derivatives in a PDE. For the linear, constant-coefficient systems of ODEs associated with our model equations, which are in the form (2.36), the solution can be written in the form (2.44), which we rewrite here as follows, noting that $t = nh$:

$$\vec{u}(t) = c_1\left(\mathrm{e}^{\lambda_1 h}\right)^n \vec{x}_1 + \cdots + c_m\left(\mathrm{e}^{\lambda_m h}\right)^n \vec{x}_m + \cdots + c_M\left(\mathrm{e}^{\lambda_M h}\right)^n \vec{x}_M \; + \; P.S., \tag{2.141}$$

where the λ_m and $\vec{x}_m$ are the eigenvalues and eigenvectors of the A matrix in the ODE system, and for the present we are not interested in the form of the particular solution ($P.S.$).

Both the explicit Euler and MacCormack methods are one-root methods; they produce one σ-root for each λ-root. If we use such a method to time march the system of ODEs, the solution of the resulting OΔEs is

$$\vec{u}_n = c_1(\sigma_1)^n\,\vec{x}_1 + \cdots + c_m(\sigma_m)^n\,\vec{x}_m + \cdots + c_M(\sigma_M)^n\,\vec{x}_M \;+\; P.S., \quad (2.142)$$

where the c_m and the $\vec{x}_m$ in the two equations are identical, and σ_m is an approximation to $\mathrm{e}^{\lambda h}$ that depends on the specific time-marching method. If the method produces one or more spurious σ-roots for each λ, as in our example of the leapfrog method, then the OΔE solution is

$$\begin{aligned}\vec{u}_n = {} & c_{11}(\sigma_1)_1^n\,\vec{x}_1 + \cdots + c_{m1}(\sigma_m)_1^n\,\vec{x}_m + \cdots + c_{M1}(\sigma_M)_1^n\,\vec{x}_M + P.S.\\ & + c_{12}(\sigma_1)_2^n\,\vec{x}_1 + \cdots + c_{m2}(\sigma_m)_2^n\,\vec{x}_m + \cdots + c_{M2}(\sigma_M)_2^n\,\vec{x}_M\\ & + c_{13}(\sigma_1)_3^n\,\vec{x}_1 + \cdots + c_{m3}(\sigma_m)_3^n\,\vec{x}_m + \cdots + c_{M3}(\sigma_M)_3^n\,\vec{x}_M\\ & + \text{etc., if there are more spurious roots.}\end{aligned} \quad (2.143)$$

The σ-root that approximates $\mathrm{e}^{\lambda_m h}$ is referred to as the *principal* σ-root, and designated $(\sigma_m)_1$. Application of the same time-marching method to all of the equations in a coupled system of linear ODEs in the form of (2.36) always produces one principal σ-root for every λ-root that satisfies the relation

$$\sigma = 1 + \lambda h + \frac{1}{2}\lambda^2 h^2 + \cdots + \frac{1}{k!}\lambda^k h^k + O\left(h^{k+1}\right), \quad (2.144)$$

where k is the order of the time-marching method. This property can be stated regardless of the details of the time-marching method, knowing only that its leading error is $O\left(h^{k+1}\right)$. Thus the principal root is an approximation to $\mathrm{e}^{\lambda h}$ up to $O\left(h^k\right)$.

Spurious roots arise if a method uses data from time level $n-1$ or earlier to advance the solution from time level n to $n+1$. Such roots originate entirely from the numerical approximation of the time-marching method and have nothing to do with the ODE being solved. However, generation of spurious roots does not, in itself, make a method inferior. In fact, many very accurate methods in practical use for integrating some forms of ODEs have spurious roots. Based on the starting technique, the magnitudes of the coefficients of the spurious roots will be small but nonzero. If the spurious roots themselves have amplitudes less than unity, they will not grow and hence will not contaminate the solution. Thus while spurious roots must be considered in *stability* analysis, they play virtually no role in *accuracy* analysis. Table 2.1 shows the λ–σ relations for various methods.

2.6.3 Implementation of Implicit Methods

Although the approach we have presented for analyzing time-marching methods based on the representative ODE is a powerful means of understanding the behaviour of time-marching methods, it obscures some aspects of the implementation of implicit methods to systems of nonlinear ODEs. These are introduced here.

Table 2.1 Some λ–σ relations

1.	$\sigma - 1 - \lambda h = 0$	Explicit Euler
2.	$\sigma^2 - 2\lambda h \sigma - 1 = 0$	Leapfrog
3.	$\sigma^2 - (1 + \frac{3}{2}\lambda h)\sigma + \frac{1}{2}\lambda h = 0$	AB2
4.	$\sigma^3 - (1 + \frac{23}{12}\lambda h)\sigma^2 + \frac{16}{12}\lambda h \sigma - \frac{5}{12}\lambda h = 0$	AB3
5.	$\sigma^{(}1 - \lambda h) - 1 = 0$	Implicit Euler
6.	$\sigma(1 - \frac{1}{2}\lambda h) - (1 + \frac{1}{2}\lambda h) = 0$	Trapezoidal
7.	$\sigma^2(1 - \frac{2}{3}\lambda h) - \frac{4}{3}\sigma + \frac{1}{3} = 0$	2nd-order backward
8.	$\sigma^2(1 - \frac{5}{12}\lambda h) - (1 + \frac{8}{12}\lambda h)\sigma + \frac{1}{12}\lambda h = 0$	AM3
9.	$\sigma^2 - (1 + \frac{13}{12}\lambda h + \frac{15}{24}\lambda^2 h^2)\sigma + \frac{1}{12}\lambda h(1 + \frac{5}{2}\lambda h) = 0$	ABM3
10.	$\sigma^3 - (1 + 2\lambda h)\sigma^2 + \frac{3}{2}\lambda h \sigma - \frac{1}{2}\lambda h = 0$	Gazdag
11.	$\sigma - 1 - \lambda h - \frac{1}{2}\lambda^2 h^2 = 0$	RK2
12.	$\sigma - 1 - \lambda h - \frac{1}{2}\lambda^2 h^2 - \frac{1}{6}\lambda^3 h^3 - \frac{1}{24}\lambda^4 h^4 = 0$	RK4
13.	$\sigma^2(1 - \frac{1}{3}\lambda h) - \frac{4}{3}\lambda h \sigma - (1 + \frac{1}{3}\lambda h) = 0$	Milne 4th

Application to Systems of Equations. Consider the application of the implicit Euler method to our generic system of equations given by

$$\vec{u}' = A\vec{u} - \vec{f}(t), \tag{2.145}$$

where $\vec{u}$ and $\vec{f}$ are vectors, and we still assume that A is not a function of $\vec{u}$ or t. One obtains the following system of algebraic equations that must be solved at each time step:

$$(I - hA)\vec{u}_{n+1} - \vec{u}_n = -h\vec{f}(t + h) \tag{2.146}$$

or

$$\vec{u}_{n+1} = (I - hA)^{-1}[\vec{u}_n - h\vec{f}(t + h)]. \tag{2.147}$$

The inverse is not actually performed; rather we solve (2.146) as a linear system of equations. For our one-dimensional examples, the system of equations which must be solved is tridiagonal (e.g. for periodic convection, $A = -aB_p(-1, 0, 1)/2\Delta x$), and hence its solution is inexpensive, but in multiple dimensions the bandwidth can be very large. In general, the cost per time step of an implicit method is thus larger than that of an explicit method. The primary area of application of implicit methods is in the solution of *stiff* ODEs; this is further discussed in Sect. 2.7.

Application to Nonlinear Equations. Now consider the general *nonlinear* scalar ODE given by

$$\frac{\mathrm{d}u}{\mathrm{d}t} = F(u, t). \tag{2.148}$$

Application of the implicit Euler method gives

$$u_{n+1} = u_n + hF(u_{n+1}, t_{n+1}). \tag{2.149}$$

This is a nonlinear difference equation which requires a nontrivial method to solve for u_{n+1}. There are several different approaches one can take to solving this nonlinear difference equation. An iterative method, such as Newton's method, can be used. Other alternatives include *local linearization* and *dual time stepping*.

In order to implement a local linearization, we expand $F(u,t)$ about some reference point in time. Designate the reference value by t_n and the corresponding value of the dependent variable by u_n. A Taylor series expansion about these reference quantities gives

$$F(u,t) = F_n + \left(\frac{\partial F}{\partial u}\right)(u - u_n) + \left(\frac{\partial F}{\partial t}\right)(t - t_n) + O(h^2). \tag{2.150}$$

This represents a second-order, locally-linear approximation to $F(u,t)$ that is valid in the vicinity of the reference station t_n and the corresponding $u_n = u(t_n)$. With this we obtain the locally (in the neighborhood of t_n) linear representation of (2.148), namely

$$\frac{\mathrm{d}u}{\mathrm{d}t} = \left(\frac{\partial F}{\partial u}\right)_n u + \left[F_n - \left(\frac{\partial F}{\partial u}\right)_n u_n\right] + \left(\frac{\partial F}{\partial t}\right)_n (t - t_n) + O(h^2). \tag{2.151}$$

As an example of how such an expansion can be used, consider the mechanics of applying the trapezoidal method for the time integration of (2.148). The trapezoidal method is given by

$$u_{n+1} = u_n + \frac{1}{2}h(F_{n+1} + F_n). \tag{2.152}$$

Using (2.150) to evaluate $F_{n+1} = F(u_{n+1}, t_{n+1})$, one finds

$$u_{n+1} = u_n + \frac{1}{2}h\left[F_n + \left(\frac{\partial F}{\partial u}\right)_n (u_{n+1} - u_n) + h\left(\frac{\partial F}{\partial t}\right)_n + O(h^2) + F_n\right]. \tag{2.153}$$

Note that the $O(h^2)$ term within the brackets (which is due to the local linearization) is multiplied by h and therefore preserves the second-order accuracy of the trapezoidal method. The local time linearization updated at the end of each time step and the trapezoidal time-marching method combine to make a *second-order-accurate* numerical integration process. There are, of course, other second-order implicit time-marching methods that can be used. The important point to be made here is that local linearization updated at each time step has not reduced the order of accuracy of a second-order time-marching process. Extension to systems of equations is straightforward, with $\left(\frac{\partial F}{\partial u}\right)_n$ representing a Jacobian matrix.

A useful reordering of the terms in (2.153) results in the expression

$$\left[1 - \frac{1}{2}h\left(\frac{\partial F}{\partial u}\right)_n\right]\Delta u_n = hF_n + \frac{1}{2}h^2\left(\frac{\partial F}{\partial t}\right)_n, \tag{2.154}$$

which is known as the *delta form*. In many fluid mechanics applications the nonlinear function F is not an *explicit* function of t. In such cases the partial derivative of $F(u)$ with respect to t is zero, and (2.154) simplifies to the second-order-accurate expression

$$\left[1 - \frac{1}{2}h\left(\frac{\partial F}{\partial u}\right)_n\right]\Delta u_n = hF_n. \tag{2.155}$$

Following the same steps with the implicit Euler method and again assuming that F is not an explicit function of time, we arrive at the form

$$\left[1 - h\left(\frac{\partial F}{\partial u}\right)_n\right]\Delta u_n = hF_n. \tag{2.156}$$

We see that the only difference between the implementation of the trapezoidal method and the implicit Euler method is the factor of $1/2$ in the brackets of the left side of (2.155) and (2.156). While a method of second-order accuracy or higher is preferred for unsteady problems, the first-order implicit Euler method is an excellent choice for steady problems.

Consider the limit $h \to \infty$ of (2.156) obtained by dividing both sides by h and setting $1/h = 0$. There results

$$-\left(\frac{\partial F}{\partial u}\right)_n \Delta u_n = F_n \tag{2.157}$$

or

$$u_{n+1} = u_n - \left[\left(\frac{\partial F}{\partial u}\right)_n\right]^{-1} F_n. \tag{2.158}$$

This is the well-known Newton method for finding the roots of the nonlinear equation $F(u) = 0$.

Finally, we illustrate the dual time-stepping approach by applying it to the trapezoidal method. The algebraic equation that must be solved at each time step is given by (2.152). Hence u_{n+1} is the solution to

$$G(u) = 0, \tag{2.159}$$

where

$$G(u) = -u + u_n + \frac{1}{2}h\left(F(u) + F(U_n)\right). \tag{2.160}$$

While Newton's method provides one option for solving such an equation, another approach is to consider u_{n+1} to be the steady solution of the following ODE:

$$\frac{\mathrm{d}u}{\mathrm{d}\tau} = G(u), \tag{2.161}$$

where τ is often referred to as pseudo time. One can use an appropriate time-marching method to solve this ODE, and typically the method would be optimized for obtaining steady solutions efficiently. Note that $\Delta\tau$ can be selected for rapid convergence to the steady solution of (2.161), while h determines the time accuracy of the trapezoidal method. If an *explicit* time-marching method is used to solve (2.161) for a system of equations, then one has an implementation of an implicit method that does not require the solution of a linear system of algebraic equations at each time step.

2.7 Stability Analysis

Stability of numerical algorithms for the solution of PDEs is an important and complex topic. Here we will simplify matters and consider only time-dependent ODEs and OΔEs in which the coefficient matrices are independent of both u and t. We will refer to such matrices as *stationary*. In the preceding sections, we developed the representative forms of ODEs generated from the basic PDEs by the semi-discrete approach, and then the OΔEs generated from the representative ODEs by application of time-marching methods. These are represented by

$$\frac{\mathrm{d}\vec{u}}{\mathrm{d}t} = A\vec{u} - \vec{f}(t) \tag{2.162}$$

and

$$\vec{u}_{n+1} = C\vec{u}_n - \vec{g}_n, \tag{2.163}$$

respectively. For a one-step method, the latter form is obtained by applying a time-marching method to the generic ODE form in a fairly straightforward manner. For example, the explicit Euler method leads to $C = I + hA$, and $\vec{g}_n = h\vec{f}(nh)$. Methods involving two or more steps can always be written in the form of (2.163) by introducing new dependent variables. Note also that only methods in which the time and space discretizations are treated separately can be written in an intermediate semi-discrete form such as (2.162). The fully-discrete form, (2.163), and the associated stability definitions and analysis are applicable to all methods.

Our definitions of stability are based entirely on the behavior of the homogeneous parts of (2.162) and (2.163). The stability of (2.162) depends entirely on the eigensystem[10] of A. The stability of (2.163) can often also be related to the eigensystem of its matrix. However, in this case the situation is not quite so simple since, in our applications to partial differential equations (especially hyperbolic ones), a stability definition can depend on both the time and space differencing. Analysis of these eigensystems has the important added advantage that it gives an estimate of the *rate* at which a solution approaches a steady-state if a system is stable. We will consider only systems with complete eigensystems; for a discussion of defective systems, see Lomax et al. [1]. Note that a complete system can be arbitrarily close to a defective one, in which case practical applications can make the properties of the latter appear to dominate.

If A and C are stationary, we can estimate their fundamental properties. For example, in Sect. 2.3.4, we found from our model ODEs for diffusion and periodic convection what could be expected for the eigenvalue spectra of practical physical problems containing these phenomena. They are important enough to be summarized by the following:

- For diffusion-dominated flows, the λ-eigenvalues tend to lie along the negative real axis.
- For convection-dominated flows, the λ-eigenvalues tend to lie along the imaginary axis.

2.7.1 Inherent Stability of ODEs

Here we state the standard stability criterion used for ordinary differential equations:

$$\text{For a } \textit{stationary} \text{ matrix } A, \text{ (2.162) is } \textit{inherently stable} \text{ if,} \\ \text{when } \vec{f} \text{ is constant, } \vec{u} \text{ remains bounded as } t \to \infty. \tag{2.164}$$

Note that inherent stability depends only on the transient solution of the ODEs.

If a matrix has a complete eigensystem, all of its eigenvectors are linearly independent, and the matrix can be diagonalized by a similarity transformation. In such a case it follows at once from (2.141), for example, that the ODEs are inherently stable if and only if

$$\Re(\lambda_m) \le 0 \quad \text{for all } m. \tag{2.165}$$

This states that, for inherent stability, all of the λ eigenvalues must lie on, or to the left of, the imaginary axis in the complex λ plane. This criterion is satisfied for the model ODEs representing both diffusion and periodic convection.

[10] This is *not* the case if the coefficient matrix depends on t, even if it is linear.

2.7.2 Numerical Stability of OΔEs

The OΔE companion to (2.164) is:

For a *stationary* matrix C, (2.163) is *numerically stable* if, when $\vec{g}$ is constant, $\vec{u}_n$ remains bounded as $n \to \infty$. (2.166)

We see that numerical stability depends only on the transient solution of the OΔEs. This definition of stability is sometimes referred to as asymptotic or time stability.

Consider a set of OΔEs governed by a complete eigensystem. The stability criterion, according to the condition set in (2.166), follows at once from a study of (2.142) and its companion for multiple σ-roots, (2.143). Clearly, for such systems a time-marching method is numerically stable if and only if

$$\left|(\sigma_m)_k\right| \leq 1 \quad \text{for all } m \text{ and } k. \tag{2.167}$$

This condition states that, for numerical stability, all of the σ eigenvalues (both principal and spurious, if there are any) must lie on or inside the unit circle in the complex σ-plane.

The most important aspect of numerical stability occurs under conditions when:

- one has inherently stable, coupled systems with λ-eigenvalues having widely separated magnitudes,

or

- we seek only to find a steady-state solution using a path that includes the unwanted transient.

In both of these cases there exist in the eigensystems relatively large values of $|\lambda h|$ associated with eigenvectors that we wish to drive through the solution process without any regard for their individual accuracy. This situation is the major motivation for the study of numerical stability and leads to the subject of stiffness discussed later in this section.

2.7.3 Unconditional Stability, A-stable Methods

A numerical method is *unconditionally stable* if it is stable for all ODEs that are inherently stable. A method with this property is said to be *A-stable*. It can be proved that the order of an A-stable linear multistep method *cannot exceed two*, and, furthermore that of all 2nd-order A-stable methods, the trapezoidal method has the smallest truncation error.

2.7.4 Stability Contours in the Complex λh Plane.

A convenient way to present the stability properties of a time-marching method is to plot the locus of the complex λh for which $|\sigma| = 1$, such that the resulting contour goes through the point $\lambda h = 0$. Here $|\sigma|$ refers to the maximum absolute value of any σ, principal or spurious, that is a root to the characteristic polynomial for a given λh. It follows from Sect. 2.7.2 that on one side of this contour the numerical method is stable, while on the other, it is unstable. We refer to it, therefore, as a *stability contour*.

Consider, for example, the explicit Euler method, for which

$$\sigma = 1 + \lambda h = 1 + \lambda_r h + \mathrm{i}\lambda_i h, \tag{2.168}$$

where λ_r and λ_i denote the real and imaginary parts of λ. Setting $|\sigma| = 1$ leads to

$$(1 + \lambda_r h)^2 + (\lambda_i h)^2 = 1, \tag{2.169}$$

which is the equation of a unit circle in the complex λh plane centered at $(-1, 0)$. The explicit Euler method is stable for λh values on or inside this circle. This means that it is unstable for the model periodic convection ODE and convection-dominated problems in general. For the model diffusion ODE it is *conditionally stable*. The time step must be chosen such that the eigenvalue of largest magnitude, which is given by

$$\lambda = \frac{\nu}{\Delta x^2}\left[-2 + 2\cos\left(\frac{M\pi}{M+1}\right)\right] \tag{2.170}$$

lies on or inside the unit circle. This gives

$$h \leq \frac{\Delta x^2}{\nu\left[1 - \cos\left(\frac{M\pi}{M+1}\right)\right]} \approx \frac{\Delta x^2}{2\nu}, \tag{2.171}$$

or

$$\frac{\nu h}{\Delta x^2} \leq \frac{1}{2}, \tag{2.172}$$

where $\nu h/\Delta x^2$ is often referred to as the *Von Neumann number*.

The stability contour of the explicit Euler method is typical of all stability contours for explicit methods in the following two ways:

(1) The contour encloses a finite portion of the left-half complex λh-plane.
(2) The region of stability is *inside* the boundary, and therefore, it is conditional.

Fig. 2.5 Stability contours for explicit Runge–Kutta methods

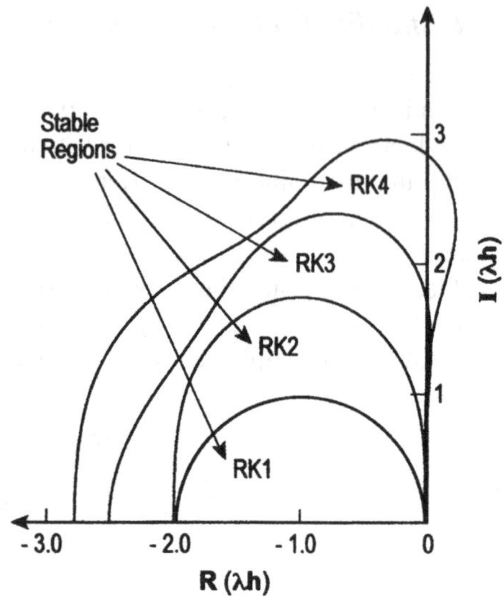

Stability contours for explicit Runge-Kutta methods of orders one through four are shown in Fig. 2.5.[11] Notice that the contours of the third- and fourth-order Runge–Kutta methods include a portion of the imaginary axis out to ± 1.9i and $\pm 2\sqrt{2}$i, respectively, and hence are suitable for convection-dominated problems.

The eigenvalues of the ODE system arising from the application of second-order centered differencing to the periodic convection PDE are given in (2.46). The maximum magnitude is $|a|/\Delta x$, which leads to the following time step restriction when this system is solved using the fourth-order Runge-Kutta method:

$$\frac{|a|h}{\Delta x} \leq 2\sqrt{2}, \tag{2.173}$$

where $|a|h/\Delta x$ is known as the *Courant* or *CFL number*.

For the implicit Euler method, one can easily show that the stability contour is a unit circle centered at $(1, 0)$ with the unstable region being *inside* the circle. This means that the method is numerically stable even when the ODEs that it is being used to integrate are inherently unstable and is typical of many stability contours for unconditionally stable implicit methods. For the trapezoidal method, the stability boundary is the imaginary axis, so it is stable for λh lying on or to the left of this axis. Hence its stability condition precisely mimics that of the ODE system.

[11] The method labelled RK1 is the explicit Euler method.

2.7.5 Fourier Stability Analysis

The most popular form of stability analysis for numerical schemes is the Fourier or Von Neumann approach. This analysis is usually carried out on point operators, and it does not depend on an intermediate stage of ODEs. Strictly speaking it applies only to difference approximations of PDEs that produce OΔEs which are linear, have no space or time varying coefficients, and have periodic boundary conditions. In practical application it is often used as a guide for estimating the worthiness of a method for more general problems. It serves as a fairly reliable *necessary* stability condition, but it is by no means a *sufficient* one.

One takes data from a "typical" point in the flow field and uses this as constant throughout time and space according to the assumptions given above. Then one imposes a spatial harmonic as an initial value on the mesh and asks the question: Will its amplitude grow or decay in time? The answer is determined by finding the conditions under which

$$u(x, t) = \mathrm{e}^{\alpha t} \cdot \mathrm{e}^{\mathrm{i}\kappa x} \tag{2.174}$$

is a solution to the *difference* equation, where κ is real and $\kappa \Delta x$ lies in the range $0 \leq \kappa \Delta x \leq \pi$. Since, for the general term,

$$u_{j+m}^{(n+\ell)} = \mathrm{e}^{\alpha(t+\ell\Delta t)} \cdot \mathrm{e}^{\mathrm{i}\kappa(x+m\Delta x)} = \mathrm{e}^{\alpha\ell\Delta t} \cdot \mathrm{e}^{\mathrm{i}\kappa m\Delta x} \cdot u_j^{(n)},$$

the quantity $u_j^{(n)}$ is common to every term and can be factored out. In the remaining expressions, we find the term $\mathrm{e}^{\alpha\Delta t}$, which we represent by σ, thus

$$\sigma \equiv \mathrm{e}^{\alpha\Delta t}.$$

Then, since $\mathrm{e}^{\alpha t} = \left(\mathrm{e}^{\alpha\Delta t}\right)^n = \sigma^n$, it is clear that:

$$\text{For numerical stability } |\sigma| \leq 1 \tag{2.175}$$

and the problem is to solve for the σs produced by any given method and, as a necessary condition for stability, make sure that, in the worst possible combination of parameters, (2.175) is satisfied.

The procedure can best be explained by an example. Consider the following fully-discrete point operator for the model diffusion equation:

$$u_j^{(n+1)} = u_j^{(n-1)} + \nu\frac{2\Delta t}{\Delta x^2}\left(u_{j+1}^{(n)} - 2u_j^{(n)} + u_{j-1}^{(n)}\right), \tag{2.176}$$

which is obtained by combining second-order centered differencing with the leapfrog method. Substitution of (2.174) into (2.176) gives the relation

$$\sigma = \sigma^{-1} + \nu\frac{2\Delta t}{\Delta x^2}\left(e^{i\kappa\Delta x} - 2 + e^{-i\kappa\Delta x}\right)$$

or

$$\sigma^2 + \underbrace{\left[\frac{4\nu\Delta t}{\Delta x^2}(1 - \cos\kappa\Delta x)\right]}_{2b}\sigma - 1 = 0. \tag{2.177}$$

Thus (2.174) is a solution of (2.176) if σ is a root of (2.177). The two roots of (2.177) are

$$\sigma_{1,2} = -b \pm \sqrt{b^2 + 1},$$

from which it is clear that one $|\sigma|$ is always > 1. We find, therefore, that by the Fourier stability test, this method is unstable for all ν, κ and Δt. The same conclusion can be gleaned from a knowledge of the stability contour for the leapfrog method and the eigenvalues of the diffusion ODE system. The leapfrog method is stable only for pure imaginary eigenvalues with amplitude less than or equal to unity. The diffusion ODE eigenvalues are strictly real and hence cannot be brought into the stable region of the leapfrog method by any choice of h.

2.7.6 Stiffness of Systems of ODEs

The concept referred to as "stiffness" comes about from the numerical analysis of mathematical models constructed to simulate dynamic phenomena containing widely different time scales. The difference between the dynamic scales translates into a difference in the magnitudes of the eigenvalues of the ODE system. The concept of stiffness in CFD arises from the fact that we often do not need accurate *time resolution* of eigenvectors associated with the large $|\lambda_m|$ in the transient solution, although these eigenvectors must remain coupled into the system to maintain the accuracy of the *spatial resolution*. For example, recall the modified wavenumber for a second-order centered difference approximation of a first derivative depicted in Fig. 2.2. For wavenumbers $\kappa\Delta x$ greater than unity the approximation is very inaccurate. Therefore there is no reason to time-march the eigenvectors associated with these wavenumbers with a high degree of accuracy. However, these components of the solution must be time-marched in a stable manner so that they do not contaminate the solution.

This situation is depicted graphically in Fig. 2.6 for the explicit Euler method. All eigenvalues, whether they must be accurately time resolved or not, must lie within the stable region of the time-marching method. In addition, those eigenvalues that correspond to eigenvectors for which accurate time resolution is required must lie within a region near the origin where the principal σ-root is a sufficiently accurate approximation to $e^{\lambda h}$ for the purposes of the specific simulation (labelled the accurate

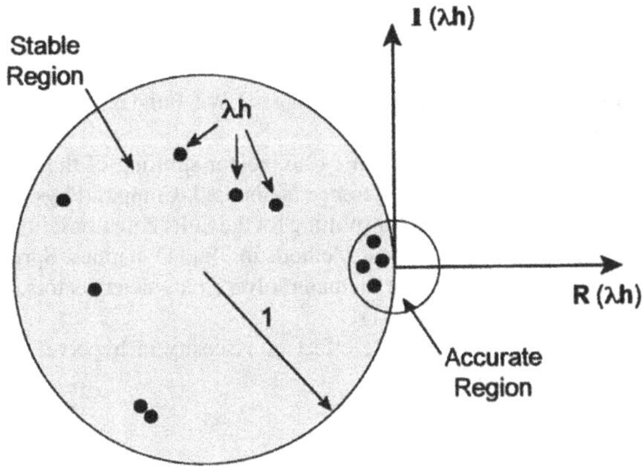

Fig. 2.6 Stable and accurate regions for the explicit Euler method

region). In the figure, the time step has been chosen so that time accuracy is given to the eigenvectors associated with the eigenvalues lying in the small circle, and stability without time accuracy is given to those associated with the eigenvalues lying outside of the small circle but still inside the large circle.

We term the eigenvalues corresponding to eigenvectors for which time accuracy is required the *driving* eigenvalues, and those for which only stability is required are termed *parasitic* eigenvalues. Unfortunately, although time accuracy requirements are dictated by the driving eigenvalues, numerical stability requirements are dictated by the parasitic ones. If a time step h chosen on the basis of stability requirements is sufficiently small that the driving eigenvalues fall within the accurate region, then the time step choice is described as *stability limited.* Similarly, if a time step h chosen on the basis of accuracy requirements is sufficiently small that the parasitic eigenvalues fall within the stable region, then the time step choice is described as *accuracy limited.*

The stiffness of an ODE system is related to the ratio of the magnitude of the largest parasitic eigenvalue to that of the largest driving eigenvalue. If this ratio is large, the system is stiff, and, if a conditionally stable time-marching method is used, the time step selection will be severely constrained by stability requirements. In other words, the time step necessary for stability is much smaller than that required for accuracy of the driving eigenvalues, and the simulation can be inefficient, requiring many more time steps than are actually needed for accurate resolution of the driving modes. In such instances, unconditionally stable implicit methods become preferable, as the time step can be selected solely on the basis of the accuracy requirements. Since implicit methods typically require more computation per time step, the comparison depends on the degree of stiffness of the problem. As the degree of stiffness increases, the advantage tilts toward implicit methods, as the reduced number of time steps begins to outweigh the increased cost per time step.

References

1. Lomax, H., Pulliam, T.H., Zingg, D.W.: Fundamentals of Computational Fluid Dynamics. Springer, Berlin (2001)
2. Steger, J.L., Warming, R.F.: Flux vector splitting of the inviscid gas dynamic equations with applications to finite difference methods. J. Comput. Phys. **40**, 263–293 (1981)
3. Van Leer, B.: Flux vector splitting for the Euler equations. In: Proceedings of the 8th International Conference on Numerical Methods in Fluid Dynamics, Springer-Verlag, Berlin (1982)
4. Roe, P.L.: Approximate Riemann solvers, parameter vectors, and difference schemes. J. Comput. Phys. **43**, 357–372 (1981)
5. MacCormack, R.W.: The effect of viscosity in hypervelocity impact cratering, AIAA Paper 69-354 (1969)

Chapter 3
Governing Equations

The governing equations are presented in the PDE form solved numerically by finite-difference methods as well as the integral form solved numerically by finite-volume methods. In addition, the quasi-one-dimensional Euler equations and the shock-tube problem are given, along with a means for obtaining their exact solutions. These form the basis of the programming assignments in this and subsequent chapters.

3.1 The Euler and Navier-Stokes Equations

3.1.1 Partial Differential Equation Form

Flow of a continuum fluid is governed by a set of partial differential equations collectively known as the Navier-Stokes equations.[1] They can be written in various different forms. We present the following form, known as conservative form, because it is advantageous for numerical solution, as we shall see later, and restrict our interest to two-dimensional Cartesian coordinates for simplicity of exposition. Extension to three dimensions is straightforward. In two dimensions, there are four equations, representing the conservation of mass, two components of momentum, and energy. For an unsteady compressible flow, these can be written as follows:

$$\frac{\partial Q}{\partial t} + \frac{\partial E}{\partial x} + \frac{\partial F}{\partial y} = \frac{\partial E_{\mathrm{v}}}{\partial x} + \frac{\partial F_{\mathrm{v}}}{\partial y}, \tag{3.1}$$

[1] Formally, the Navier-Stokes equations are the equations arising from the conservation of momentum; they do not include the equations describing conservation of mass and energy. We follow the prevailing usage and term the whole set the Navier-Stokes equations.

T. H. Pulliam and D. W. Zingg, *Fundamental Algorithms in Computational Fluid Dynamics*, Scientific Computation, DOI: 10.1007/978-3-319-05053-9_3,

where

$$Q = \begin{bmatrix} \rho \\ \rho u \\ \rho v \\ e \end{bmatrix}, \quad E = \begin{bmatrix} \rho u \\ \rho u^2 + p \\ \rho uv \\ u(e+p) \end{bmatrix}, \quad F = \begin{bmatrix} \rho v \\ \rho uv \\ \rho v^2 + p \\ v(e+p) \end{bmatrix}, \tag{3.2}$$

$$E_{\rm v} = \begin{bmatrix} 0 \\ \tau_{xx} \\ \tau_{xy} \\ f_4 \end{bmatrix}, \quad F_{\rm v} = \begin{bmatrix} 0 \\ \tau_{xy} \\ \tau_{yy} \\ g_4 \end{bmatrix}. \tag{3.3}$$

The variable Q represents the conservative dependent variables per unit volume, including the density, ρ, the components of momentum per unit volume, ρu and ρv, where u and v are the Cartesian velocity components, and the total energy per unit volume, e. The total energy includes internal and kinetic energy and can be written as

$$e = \rho\left(\epsilon + \frac{u^2+v^2}{2}\right), \tag{3.4}$$

where ϵ is the internal energy per unit mass. The vectors E and F are known as the inviscid flux vectors. They contain convective fluxes plus terms associated with pressure. For some flow problems, other terms, such as gravitational forces, can be important and should be included. In the momentum equations, the pressure terms represent forces; in the energy equation they are associated with the work done by the pressure forces. Although it is important to understand these equations as conservation laws for mass, momentum, and energy, it is also instructive to recognize that the momentum equations are an expression of the fact that in an inertial reference frame, the time rate of change of momentum of a particle or collection of particles is equal to the net force acting on the particle or collection of particles. In other words, the momentum equations are a statement of Newton's second law, force equals mass times acceleration.

We will restrict our attention here to thermally and calorically perfect gases, giving the relations

$$p = \rho RT \tag{3.5}$$

and

$$\epsilon = c_{\rm v} T, \tag{3.6}$$

where p is the pressure, R is the specific gas constant, T is the temperature, and $c_{\rm v}$ is the specific heat capacity at constant volume. The equation of state enables the pressure to be expressed in terms of the conservative flow variables as follows:

$$p = \rho RT \tag{3.7}$$

$$= \rho R \left(\frac{\epsilon}{c_{\mathrm{v}}} \right) \tag{3.8}$$

$$= (\gamma - 1)\rho\epsilon \tag{3.9}$$

$$= (\gamma - 1)\left(e - \frac{\rho}{2}(u^2 + v^2) \right) \tag{3.10}$$

$$= (\gamma - 1)\left[e - \frac{1}{2\rho}\left((\rho u)^2 + (\rho v)^2 \right) \right], \tag{3.11}$$

where γ is the ratio of specific heats, $c_{\mathrm{p}}/c_{\mathrm{v}}$, c_{p} is the specific heat capacity at constant pressure, and we have used the relation

$$c_{\mathrm{v}} = \frac{R}{\gamma - 1}. \tag{3.12}$$

For a perfect gas, the speed of sound, a, satisfies the relations

$$a^2 = \frac{\gamma p}{\rho} = \gamma R T. \tag{3.13}$$

Alternative equations of state must be used under conditions when the perfect gas law does not apply, such as flows at very high temperatures.

The vectors E_{v} and F_{v} include terms associated with viscosity and heat conduction. We will consider Newtonian fluids here, but the reader is reminded that this assumption is not universally applicable. For a Newtonian fluid, the viscous stresses are given in two dimensions by

$$\tau_{xx} = \mu \left(\frac{4}{3}\frac{\partial u}{\partial x} - \frac{2}{3}\frac{\partial v}{\partial y} \right),$$
$$\tau_{xy} = \mu \left(\frac{\partial u}{\partial y} + \frac{\partial v}{\partial x} \right),$$
$$\tau_{yy} = \mu \left(-\frac{2}{3}\frac{\partial u}{\partial x} + \frac{4}{3}\frac{\partial v}{\partial y} \right), \tag{3.14}$$

where μ is the dynamic viscosity, which is typically a function of temperature, and for air can often be determined using Sutherland's law. The viscous terms appearing in the momentum equations are forces. The terms f_4 and g_4 in the energy equation represent the work done by the viscous forces as well as heat conduction.

Heat conduction is governed by Fourier's law, which states that the local heat flux, which is the rate of flow of heat per unit area per unit time, is directly proportional to the local gradient of the temperature. The constant of proportionality, k, is known as the thermal conductivity. Based on Fourier's law, the heat conduction terms can be written in two-dimensional Cartesian coordinates as

$$\frac{\partial}{\partial x}\left(k \frac{\partial T}{\partial x} \right) + \frac{\partial}{\partial y}\left(k \frac{\partial T}{\partial y} \right). \tag{3.15}$$

It is convenient to introduce the Prandtl number, Pr, which is the ratio of kinematic viscosity to thermal diffusivity. It is given by

$$Pr = \frac{\mu c_\mathrm{p}}{k}. \tag{3.16}$$

This dimensionless number depends on the properties of the fluid. For air, the Prandtl number is close to 0.71 for a wide range of temperatures. For a perfect gas, the heat conduction terms can thus be written as

$$\frac{\partial}{\partial x}\left(\frac{\mu}{Pr(\gamma-1)}\frac{\partial a^2}{\partial x}\right)+\frac{\partial}{\partial y}\left(\frac{\mu}{Pr(\gamma-1)}\frac{\partial a^2}{\partial y}\right), \tag{3.17}$$

where we have used the relation

$$c_\mathrm{p} = \frac{\gamma R}{\gamma - 1}. \tag{3.18}$$

Hence we obtain the following expressions for the terms f_4 and g_4 in the energy equation:

$$\begin{aligned} f_4 &= u\tau_{xx} + v\tau_{xy} + \frac{\mu}{Pr(\gamma-1)}\frac{\partial a^2}{\partial x}, \\ g_4 &= u\tau_{xy} + v\tau_{yy} + \frac{\mu}{Pr(\gamma-1)}\frac{\partial a^2}{\partial y}. \end{aligned} \tag{3.19}$$

It is often convenient to non-dimensionalize the equations. In order to do so, we require a reference length, l, normally chosen as some characteristic physical length scale in the problem, a reference density, ρ_∞, often chosen for an external flow as the density of the undisturbed fluid far from the body, and a reference velocity scale. It is traditional in fluid dynamics to choose a velocity scale such as u_∞, the velocity of the body moving through the fluid. For our purpose here, it is more convenient to use a_∞, the speed of sound in the undisturbed air far from the body, since u_∞ could be zero for some flow problems, such as a helicopter in hover. The conditions far from the body are often called free stream conditions. With these reference quantities, we obtain the following non-dimensional quantities (indicated by the tilde):

$$\begin{aligned} &\widetilde{x} = \frac{x}{l}, \quad \widetilde{y} = \frac{y}{l}, \quad \widetilde{t} = \frac{t a_\infty}{l}, \\ &\widetilde{\rho} = \frac{\rho}{\rho_\infty}, \quad \widetilde{u} = \frac{u}{a_\infty}, \quad \widetilde{v} = \frac{v}{a_\infty}, \\ &\widetilde{e} = \frac{e}{\rho_\infty a_\infty^2}, \quad \widetilde{\mu} = \frac{\mu}{\mu_\infty}. \end{aligned} \tag{3.20}$$

Substituting these non-dimensional quantities into the Navier-Stokes equations, dropping the tildes, and defining the Reynolds number as

$$Re = \frac{\rho_\infty l a_\infty}{\mu_\infty}, \tag{3.21}$$

we obtain the following non-dimensional form of the equations:

$$\frac{\partial Q}{\partial t} + \frac{\partial E}{\partial x} + \frac{\partial F}{\partial y} = Re^{-1}\left(\frac{\partial E_\mathrm{v}}{\partial x} + \frac{\partial F_\mathrm{v}}{\partial y}\right), \tag{3.22}$$

where all terms are as previously defined except in terms of non-dimensional quantities. It is important to note that this definition of the Reynolds number based on a_∞ differs from the conventional definition based on u_∞. The two are related by the free stream Mach number, $M_\infty = u_\infty/a_\infty$.

The Euler equations are obtained from the Navier-Stokes equations by neglecting the terms associated with viscosity and heat conduction, i.e. setting E_v and F_v to zero. Numerical solutions of the Euler equations can be useful if the effect of viscosity and heat conduction on the quantities of interest is small. There are many other simplified forms of the Navier-Stokes equations that can be useful for specific classes of problems. It is important that their limitations be well understood.

We stated earlier that the above equations are in conservative form. There are two aspects to this. The first is that we choose the conserved quantities, mass, momentum, and energy, per unit volume as the dependent variables. It is also possible to write a system of equations in terms of other variables, such as the *primitive* variables, density, velocity, and pressure, that is analytically equivalent but can lead to different solutions when solved numerically. For example, for a perfect gas the one-dimensional Euler equations can be written in terms of the *primitive* variables $R = [\rho, u, p]^\mathrm{T}$ as follows:

$$\frac{\partial R}{\partial t} + \tilde{A}\frac{\partial R}{\partial x} = 0\,, \tag{3.23}$$

where

$$\tilde{A} = \begin{bmatrix} u & \rho & 0 \\ 0 & u & \rho^{-1} \\ 0 & \gamma p & u \end{bmatrix}.$$

The second aspect is related to the products appearing in the fluxes. In the conservative form, the product rule of differentiation is not applied. A term such as

$$\frac{\partial}{\partial x}(\rho u)$$

appearing in the mass conservation equation is not expanded as

$$\rho \frac{\partial u}{\partial x} + u \frac{\partial \rho}{\partial x},$$

which is in non-conservative form. Again the two forms are analytically equivalent, but under some circumstances, such as flows with nonstationary shock waves, an algorithm that is not conservative can produce substantially inaccurate solutions.

Although we do not normally solve non-conservative forms of the equations, they can be useful for analysis. For example, consider the one-dimensional Euler equations in conservative form:

$$\frac{\partial Q}{\partial t} + \frac{\partial E}{\partial x} = 0, \tag{3.24}$$

where

$$Q = \begin{bmatrix} Q_1 \\ Q_2 \\ Q_3 \end{bmatrix} = \begin{bmatrix} \rho \\ \rho u \\ e \end{bmatrix}, \quad E = \begin{bmatrix} E_1 \\ E_2 \\ E_3 \end{bmatrix} = \begin{bmatrix} \rho u \\ \rho u^2 + p \\ u(e+p) \end{bmatrix}. \tag{3.25}$$

If the solution is smooth, (3.24) can be rewritten in the following form:

$$\frac{\partial Q}{\partial t} + A \frac{\partial Q}{\partial x} = 0\,, \tag{3.26}$$

where

$$A = \frac{\partial E}{\partial Q} \tag{3.27}$$

is known as the flux Jacobian. The flux Jacobian is derived by first writing the flux vector in terms of the conservative variables

$$E = \begin{bmatrix} Q_2 \\ \\ (\gamma - 1)Q_3 + \frac{3-\gamma}{2} \frac{Q_2^2}{Q_1} \\ \\ \gamma \frac{Q_3 Q_2}{Q_1} - \frac{\gamma - 1}{2} \frac{Q_2^3}{Q_1^2} \end{bmatrix}, \tag{3.28}$$

which gives, for a perfect gas,

$$A = \frac{\partial E_i}{\partial Q_j} = \begin{bmatrix} 0 & 1 & 0 \\ \frac{\gamma-3}{2} \left(\frac{Q_2}{Q_1}\right)^2 & (3-\gamma)\frac{Q_2}{Q_1} & \gamma - 1 \\ A_{31} & A_{32} & \gamma \left(\frac{Q_2}{Q_1}\right) \end{bmatrix}, \tag{3.29}$$

where

$$A_{31} = (\gamma - 1)\left(\frac{Q_2}{Q_1}\right)^3 - \gamma\left(\frac{Q_3}{Q_1}\right)\left(\frac{Q_2}{Q_1}\right)$$

$$A_{32} = \gamma\left(\frac{Q_3}{Q_1}\right) - \frac{3(\gamma - 1)}{2}\left(\frac{Q_2}{Q_1}\right)^2. \tag{3.30}$$

This can be rewritten in terms of ρ, u, and e as

$$A = \begin{bmatrix} 0 & 1 & 0 \\ \frac{\gamma-3}{2}u^2 & (3-\gamma)u & \gamma - 1 \\ A_{31} & A_{32} & \gamma u \end{bmatrix}, \tag{3.31}$$

where

$$A_{31} = (\gamma - 1)u^3 - \gamma\frac{ue}{\rho}$$

$$A_{32} = \gamma\frac{e}{\rho} - \frac{3(\gamma - 1)}{2}u^2. \tag{3.32}$$

The eigenvalues of the flux Jacobian A are u, $u + a$, $u - a$. Since these are all real, and the eigenvectors of A are linearly independent, the system (3.26) is *hyperbolic*. Hence some important properties of these equations can be obtained from characteristic theory. First, the eigenvalues represent the characteristic speeds at which information is propagated. The convection of the fluid propagates information at speed u, while sound waves propagate information at speeds $u + a$ and $u - a$. If the flow is supersonic, i.e. $|u| > a$, then all of the eigenvalues have the same sign, and information is propagated in one direction only. If the flow is subsonic, i.e. $|u| < a$, then the eigenvalues are of mixed sign, and information is propagated in both directions. This is critical in the design of numerical methods and in the development of boundary conditions. Riemann invariants can be found that are propagated at the characteristic speeds, as long as the solution remains smooth. The entropy $\ln(p/\rho^\gamma)$ propagates at speed u, while the quantities $u \pm 2a/(\gamma - 1)$ propagate at speeds $u \pm a$.

The flux Jacobian A in (3.26) is related to the matrix $\tilde{A}$ in (3.23) by the following similarity transform:

$$A = S\tilde{A}S^{-1}, \tag{3.33}$$

where $S = \partial Q/\partial R$. Hence the eigenvalues of the two matrices are identical, consistent with the fact that (3.26) and (3.23) are different representations of the same physical processes.

3.1.2 Integral Form

The Navier-Stokes equations governing an unsteady compressible flow can also be written in the following *integral* form in two-dimensional Cartesian coordinates:

$$\frac{\mathrm{d}}{\mathrm{d}t}\iint_{V(t)} Q\mathrm{d}x\mathrm{d}y + \oint_{S(t)} (E\mathrm{d}y - F\mathrm{d}x) = Re^{-1}\oint_{S(t)} (E_{\mathrm{v}}\mathrm{d}y - F_{\mathrm{v}}\mathrm{d}x) \quad (3.34)$$

for an arbitrary control volume $V(t)$ bounded by the surface $S(t)$, with all variables as defined and non-dimensionalized previously. This form is obtained from the more general coordinate-free form

$$\frac{\mathrm{d}}{\mathrm{d}t}\int_{V(t)} Q\mathrm{d}V + \oint_{S(t)} \hat{n}\cdot\mathcal{F}\mathrm{d}S = 0, \quad (3.35)$$

where $\hat{n}$ is the unit vector normal to the surface pointing outwards, and $\mathcal{F}$ is the flux tensor, including inviscid, viscous, and heat conduction terms. In two-dimensional Cartesian coordinates, the flux tensor is given by

$$\mathcal{F} = (E - Re^{-1}E_{\mathrm{v}})\hat{i} + (F - Re^{-1}F_{\mathrm{v}})\hat{j}, \quad (3.36)$$

where $\hat{i}$ and $\hat{j}$ are unit vectors in the x and y directions, respectively. The contour in (3.34) is traversed in a counter-clockwise direction; hence the area-weighted outward normal can be written as

$$\hat{n}\mathrm{d}S = \hat{i}\mathrm{d}y - \hat{j}\mathrm{d}x. \quad (3.37)$$

3.1.3 Physical Boundary Conditions

The physical boundary conditions that must be satisfied at a rigid body surface are as follows. For an inviscid flow governed by the Euler equations, the flow must be tangent to the surface; in other words, the velocity component normal to the surface must be zero:

$$(u\hat{i} + v\hat{j})\cdot\hat{n} = 0\,. \quad (3.38)$$

For viscous flows governed by the Navier-Stokes equations, the no-slip condition must be satisfied at the surface: all components of velocity must be zero. In addition, for viscous flows, it is normally assumed that the surface is either held at a fixed temperature or is adiabatic. In the latter case, the gradient of the temperature in a direction normal to the surface is zero at the surface:

$$\nabla T\cdot\hat{n} = 0. \quad (3.39)$$

Other physical boundary conditions can vary from problem to problem. For external flow problems, there is often a requirement that as the distance from the body approaches infinity, the flow must approach its undisturbed state. This condition is usually applied at a boundary some finite distance from the body. Other problems may involve specified incoming flows.

3.2 The Reynolds-Averaged Navier-Stokes Equations

When the Navier-Stokes equations are time-averaged over a time interval that is long in comparison with the turbulent time scales but short in comparison to other physical time scales, apparent stresses known as Reynolds stresses as well as additional heat flux terms appear. It is the function of a turbulence model, which typically involves the solution of one or more partial differential equations, to furnish these additional terms and thereby to provide closure to the system. For the remainder of this book, all algorithms will be presented in the context of the Euler and Navier-Stokes equations rather than the Reynolds-averaged Navier-Stokes (RANS) equations, although these algorithms are routinely used for the RANS equations. In order to apply these algorithms to the RANS equations, the Reynolds stresses must be added to the Navier-Stokes equations in the form given by the particular turbulence model selected, and the solution algorithm must be applied to any partial differential equations associated with the turbulence model.

3.3 The Quasi-One-Dimensional Euler Equations and the Shock-Tube Problem

The quasi-one-dimensional Euler equations and the shock-tube problem are used throughout this book as examples and in the programming assignments. The quasi-one-dimensional Euler equations govern the inviscid flow in a quasi-one-dimensional channel with varying cross-sectional area per unit depth $S(x)$ and can be written as follows [1]:

$$\frac{\partial(\rho S)}{\partial t} + \frac{\partial(\rho u S)}{\partial x} = 0, \tag{3.40}$$

$$\frac{\partial(\rho u S)}{\partial t} + \frac{\partial[(\rho u^2 + p)S]}{\partial x} = p\frac{\mathrm{d}S}{\mathrm{d}x}, \tag{3.41}$$

$$\frac{\partial(eS)}{\partial t} + \frac{\partial[u(e + p)S]}{\partial x} = 0, \tag{3.42}$$

where the variables t, x, ρ, u, p, and e have the same definitions as in Sect. 3.1. These are typically solved for a steady flow in a channel with prescribed boundary conditions.

The shock-tube problem is an initial-value problem. Viscosity is again neglected, and the above equations are solved with $S(x) = 1$. The initial conditions are such that there are two initial fluid states separated by a diaphragm at $t = 0$. These are typically quiescent with different pressures and densities. Using x_0 to represent the location of the diaphragm and subscripts L and R to indicate the fluid states to the left and right of the diaphragm, the initial conditions can be written as

$$u = 0,\ p = p_L,\ \rho = \rho_L,\ x < x_0 \tag{3.43}$$

$$u = 0,\ p = p_R,\ \rho = \rho_R,\ x \geq x_0\,. \tag{3.44}$$

When the diaphragm is removed instantaneously, a flow is initiated in the direction from high pressure to low. For the example given later in the section, where $p_R < p_L$, a contact discontinuity separating the original two states propagates to the right, an expansion wave propagates to the left, and a shock wave propagates to the right at a speed higher than that of the contact surface. We assume that the process is terminated before any of these waves reach the ends of the shock tube. Hence boundary conditions are not required.

3.3.1 Exact Solution: Quasi-One-Dimensional Channel Flow

We present the equations needed to write a computer program to determine the exact solution for a quasi-one-dimensional channel flow as a reference solution for comparison with numerical solutions. The relevant theory and explanation can be found in most good gasdynamics textbooks (see Shapiro [1] for example). A problem is defined by specifying the channel area variation, $S(x)$, the total pressure and temperature at the inlet, p_{01} and T_{01}, the critical area, S^*, an indication of whether the initial Mach number is subsonic or supersonic, and a shock location, x_{shock}, if applicable. The solution is calculated by marching from inlet to outlet. At a given x location, both S and S^* are known, so the local Mach number, $M = u/a$, can be calculated from the following nonlinear equation using an iterative technique:

$$\frac{S}{S^*} = \frac{1}{M}\left[\frac{2}{\gamma+1}\left(1+\frac{\gamma-1}{2}M^2\right)\right]^{\frac{\gamma+1}{2(\gamma-1)}}. \tag{3.45}$$

A subsonic or supersonic initial Mach number guess should be used, depending on the problem specification. The temperature and pressure can then be determined from the isentropic relations:

$$T = \frac{T_{01}}{1 + \frac{\gamma-1}{2}M^2} \tag{3.46}$$

$$p = p_{01}\left(1 + \frac{\gamma-1}{2}M^2\right)^{-\left(\frac{\gamma}{\gamma-1}\right)}. \tag{3.47}$$

Other variables, such as density, velocity, and sound speed, can be calculated using the perfect gas relations and the definition of the Mach number. Once the specified location of the shock is reached, if applicable, the Rankine-Hugoniot relations are used to find the conditions downstream of the shock:

$$T_{0R} = T_{0L} \tag{3.48}$$

$$M_R^2 = \frac{2 + (\gamma - 1)M_L^2}{2\gamma M_L^2 - (\gamma - 1)} \tag{3.49}$$

$$\frac{p_R}{p_L} = \frac{2\gamma M_L^2 - (\gamma - 1)}{\gamma + 1} \tag{3.50}$$

$$\frac{p_{0R}}{p_{0L}} = \frac{\left([(\gamma+1)/2]M_L^2/\{1 + [(\gamma-1)/2]M_L^2\}\right)^{\frac{\gamma}{\gamma-1}}}{\{[2\gamma/(\gamma+1)]M_L^2 - (\gamma-1)/(\gamma+1)\}^{\frac{1}{\gamma-1}}}. \tag{3.51}$$

The density and sound speed downstream of the shock can then be found using the perfect gas relations. The value of S^* must also be recalculated to correspond to conditions downstream of the shock from:

$$S_R^* = S_L^* \frac{\rho_L^* a_L^*}{\rho_R^* a_R^*}, \tag{3.52}$$

where

$$\rho_{01} = \frac{p_{01}}{RT_{01}}$$

$$\rho_0^R = \frac{p_0^R}{RT_{01}}$$

$$a_{01} = \sqrt{\frac{\gamma p_{01}}{\rho_{01}}}$$

$$a_0^R = \sqrt{\frac{\gamma p_0^R}{\rho_0^R}}$$

$$\rho_L^* a_L^* = \rho_{01} a_{01} \left(\frac{2}{\gamma+1}\right)^{\frac{\gamma+1}{2(\gamma-1)}}$$

$$\rho_R^* a_R^* = \rho_0^R a_0^R \left(\frac{2}{\gamma+1}\right)^{\frac{\gamma+1}{2(\gamma-1)}}.$$

Fig. 3.1 Exact solution for the subsonic channel flow problem. **a** Pressure (in Pa). **b** Mach number

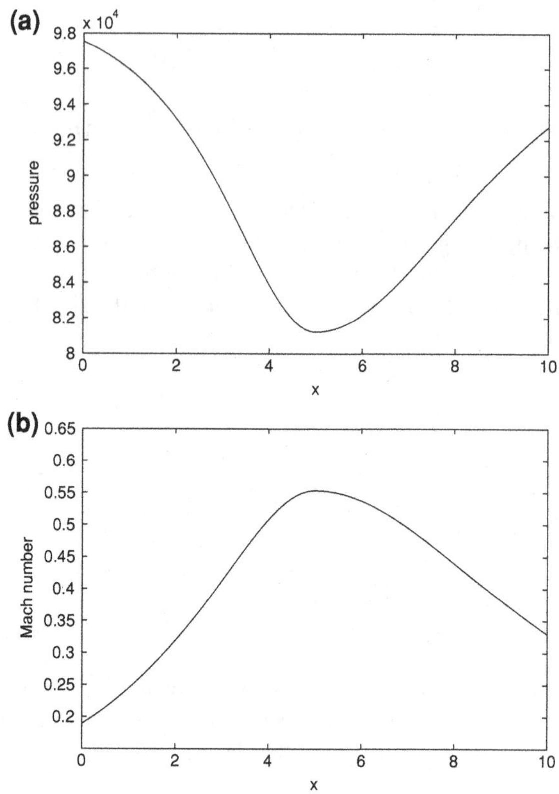

The solution downstream of the shock can then be calculated using (3.45) with these new values of S^* and p_0.

We will consider two examples from Hirsch [2]. In both cases, $S(x)$ is given by

$$S(x) = \begin{cases} 1 + 1.5\left(1 - \frac{x}{5}\right)^2 & 0 \le x \le 5 \\ 1 + 0.5\left(1 - \frac{x}{5}\right)^2 & 5 \le x \le 10 \end{cases} \tag{3.53}$$

where $S(x)$ and x are in meters. In both cases, the fluid is air, which is considered to be a perfect gas with $R = 287\ \mathrm{N \cdot m \cdot kg^{-1} \cdot K^{-1}}$, and $\gamma = 1.4$, the total temperature is $T_0 = 300$ K, and the total pressure at the inlet is $p_{01} = 100$ kPa. For the first case, the flow is subsonic throughout the channel, with $S^* = 0.8$. The pressure and Mach number for this case are plotted in Fig. 3.1. For the second case, the flow is transonic, with subsonic flow at the inlet, a shock at $x = 7$, and $S^* = 1$. The pressure and Mach number for this case are plotted in Fig. 3.2.

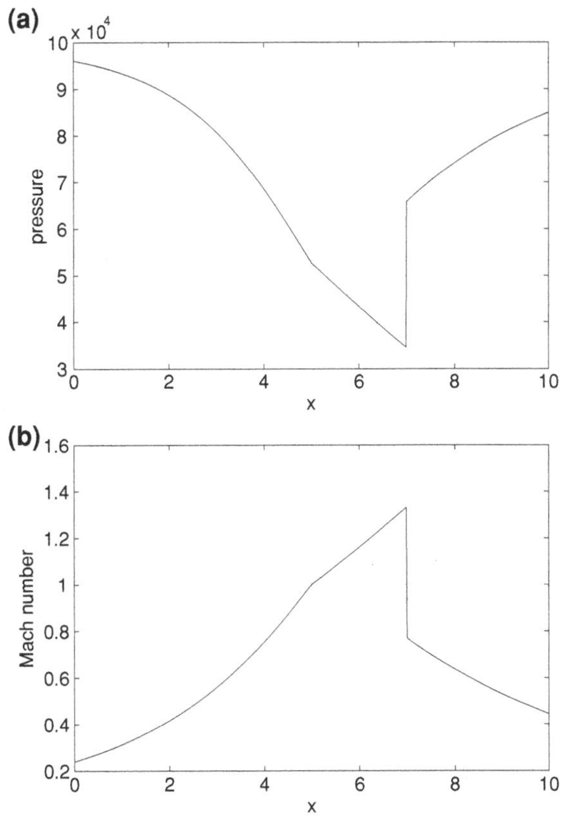

Fig. 3.2 Exact solution for the transonic channel flow problem. **a** Pressure (in Pa). **b** Mach number

3.3.2 Exact Solution: Shock-Tube Problem

As in the previous section, we present without explanation the equations needed to solve a shock-tube problem. See Hirsch [2] for more details. We assume initial conditions as described earlier in Sect. 3.3, which lead to a solution with an expansion wave traveling to the left, a contact surface moving to the right at speed V, and a shock wave moving to the right at a speed C, where $C > V$. We thus define the following states: The state to the left of the head of the expansion fan is denoted by the subscript L; it is the original quiescent state to the left of the diaphragm. The state within the expansion wave, where the variables vary continuously, is denoted by the subscript 5. The constant state between the tail of the expansion fan and the contact surface is denoted by the subscript 3. The constant state between the contact surface and the shock wave is denoted by the subscript 2. Finally, the quiescent state to the right of the shock, which is the original state to the right of the diaphragm, is denoted by the subscript R.

The normal shock relations must hold across the shock. Following Hirsch [2], we define the pressure ratio across the shock as $P = p_2/p_R$. Across the contact surface, pressure and velocity are continuous. The flow in the expansion wave is isentropic, and characteristic theory can be applied. After some algebra, the following implicit equation is found which must be solved for P:

$$\sqrt{\frac{2}{\gamma(\gamma-1)}}\frac{P-1}{\sqrt{1+\alpha P}} = \frac{2}{\gamma-1}\frac{a_L}{a_R}\left[1-\left(\frac{p_R}{p_L}P\right)^{\frac{\gamma-1}{2\gamma}}\right], \tag{3.54}$$

where

$$\alpha = \frac{\gamma+1}{\gamma-1},$$

and p_L, p_R, a_L, and a_R are the pressures and sound speeds associated with the initial states. Recall that the sound speeds can be determined from the specified pressures and densities using (3.13). Once the above equation has been solved by an iterative method for nonlinear algebraic equations, such as Newton's method, the pressure to the left of the shock, p_2, is known. The density to the left of the shock can be found from

$$\frac{\rho_2}{\rho_R} = \frac{1+\alpha P}{\alpha + P}. \tag{3.55}$$

Since the pressure is continuous across the contact surface, we know that $p_3 = p_2$. The propagation speed of the contact surface can then be found from

$$V = \frac{2}{\gamma-1}a_L\left[1-\left(\frac{p_3}{p_L}\right)^{\frac{\gamma-1}{2\gamma}}\right]. \tag{3.56}$$

The fluid velocity on either side of the contact surface must be equal to V, which gives $u_3 = u_2 = V$. To complete the state to the left of the contact surface, the density can be found by exploiting the fact that the flow in the expansion wave is isentropic, and hence the entropy to the left of the contact surface is equal to that of the original quiescent left state, giving

$$\rho_3 = \rho_L\left(\frac{p_3}{p_L}\right)^{\frac{1}{\gamma}}. \tag{3.57}$$

The speed at which the shock wave propagates is given by

$$C = \frac{(P-1)a_R^2}{\gamma u_2}. \tag{3.58}$$

The head of the expansion wave travels to the left at speed a_L. Therefore, for $x \leq x_0 - a_L t$, the fluid state is defined by the original state to the left of the diaphragm. The tail of the expansion wave moves to the left at a speed given by $a_L - V(\gamma+1)/2$. Thus the state between the tail of the expansion wave and the contact surface (state 3) is the solution for $x_0 + [V(\gamma+1)/2 - a_L]t < x \leq x_0 + Vt$. State 2 is the solution for $x_0 + Vt < x \leq x_0 + Ct$, and finally, for $x > x_0 + Ct$, the solution is the original state to the right of the diaphragm. To complete the solution, we require the state within the expansion fan, that is for $x_0 - a_L t < x \leq x_0 + [V(\gamma+1)/2 - a_L]t$. It is given by

$$
\begin{aligned}
u_5 &= \frac{2}{\gamma+1}\left(\frac{x-x_0}{t} + a_L\right) \\
a_5 &= u_5 - \frac{x-x_0}{t} \\
p_5 &= p_L\left(\frac{a_5}{a_L}\right)^{\frac{2\gamma}{\gamma-1}} \\
\rho_5 &= \frac{\gamma p_5}{a_5^2}.
\end{aligned}
$$

As an example, we consider the following shock-tube problem from Hirsch [2]: $p_L = 10^5$, $\rho_L = 1$, $p_R = 10^4$, and $\rho_R = 0.125$, where the pressures are in Pa and the densities in Kg/m^3. The fluid is a perfect gas with $\gamma = 1.4$. Figure 3.3 displays the density and Mach number at $t = 6.1$ ms. Along with the steady channel flow solutions shown in Figs. 3.1 and 3.2, this exact solution provides an excellent reference for use in verifying numerical solutions.

3.4 Exercises

3.1 Write a computer program to determine the exact solution of the quasi-one-dimensional Euler equations for the following subsonic problem. $S(x)$ is given by

$$
S(x) = \begin{cases} 1 + 1.5\left(1 - \frac{x}{5}\right)^2 & 0 \leq x \leq 5 \\ 1 + 0.5\left(1 - \frac{x}{5}\right)^2 & 5 \leq x \leq 10 \end{cases} \tag{3.59}
$$

where $S(x)$ and x are in meters. The fluid is air, which is considered to be a perfect gas with $R = 287\ \text{N}\cdot\text{m}\cdot\text{kg}^{-1}\cdot\text{K}^{-1}$, and $\gamma = 1.4$, the total temperature is $T_0 = 300$ K, and the total pressure at the inlet is $p_{01} = 100$ kPa. The flow is subsonic throughout the channel, with $S^* = 0.8$. Compare your solution with that plotted in Fig. 3.1.

3.2 Repeat Exercise 3.1 for a transonic flow in the same channel. The flow is subsonic at the inlet, there is a shock at $x = 7$, and $S^* = 1$. Compare your solution with that plotted in Fig. 3.2.

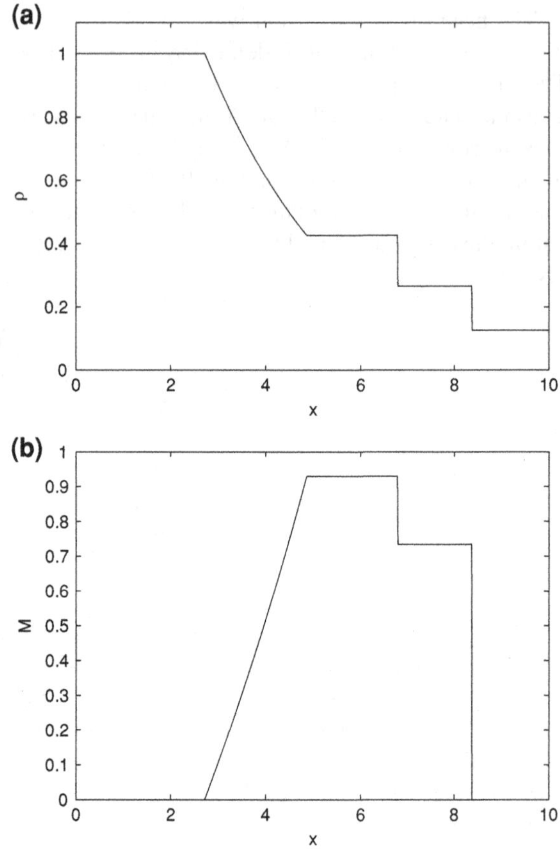

Fig. 3.3 Exact solution for the shock-tube problem at $t = 6.1$ ms. **a** Density (in Kg/m^3). **b** Mach number

3.3 Write a computer program to determine the exact solution for the following shock-tube problem: $p_L = 10^5, \rho_L = 1, p_R = 10^4$, and $\rho_R = 0.125$, where the pressures are in Pa and the densities in Kg/m^3. The fluid is a perfect gas with $\gamma = 1.4$. Compare your solution at $t = 6.1$ ms with that plotted in Fig. 3.3.

References

1. Shapiro, A.H.: The Dynamics and Thermodynamics of Compressible Fluid Flow. Ronald Press, New York (1953)
2. Hirsch, C.: Numerical Computation of Internal and External Flows, vol. 2. Wiley, Chichester (1990)

Chapter 4
An Implicit Finite-Difference Algorithm

4.1 Introduction

A numerical solution algorithm for the Navier-Stokes equations converts the original system of partial differential equations (PDEs) to a much larger system of algebraic equations, which is then solved. Many such algorithms discretize space and time independently, such that the PDEs are first reduced to ordinary differential equations (ODEs) through the discretization of the spatial terms in the governing equations. This semi-discrete ODE system is then converted to a system of ordinary difference equations (OΔEs) through a time-marching method. This assumes that the PDE system is time-dependent. If one is interested only in the steady solution of the Navier-Stokes equations, then the time-derivative terms can be dropped, and there is no intermediate ODE system. In this case, the spatial discretization directly reduces the original nonlinear PDE system to a system of nonlinear algebraic equations. Being nonlinear, this algebraic system cannot be solved directly and must be solved using an iterative method. It can often be useful to retain the time-dependent terms even if one is interested only in the steady solution, as a time-marching method that follows a quasi-physical path to the steady solution can be an effective iterative method.

Both the implicit algorithm presented in this chapter and the explicit algorithm presented in the next chapter retain the time-derivative terms in the Navier-Stokes equations even when solving for steady flows. Moreover, both algorithms involve independent discretization of space and time, and hence an intermediate semi-discrete ODE form. In principle, the spatial and temporal components of the algorithms could be presented independently. However, in these two algorithms the two are quite closely linked. In other words, the time-marching methods are particularly effective with the specific spatial discretization used. Nonetheless, the reader should be aware that it is of course possible and reasonable to develop an explicit finite-difference algorithm or an implicit finite-volume algorithm.

T. H. Pulliam and D. W. Zingg, *Fundamental Algorithms in Computational Fluid Dynamics*, Scientific Computation, DOI: 10.1007/978-3-319-05053-9_4,

The key characteristics of the algorithm presented in this chapter are as follows:

- node-based data storage; the numerical solution for the state variables is associated with the nodes of the grid
- second-order finite-difference spatial discretization; centered with added numerical dissipation; a simple shock-capturing device
- transformation to generalized curvilinear coordinates; applicable to structured grids
- implicit time marching based on approximate factorization of the resulting matrix operator

All of these terms will be explained in this chapter. Key contributions to this algorithm were made by Beam and Warming [1], Steger [2], Warming and Beam [3], Pulliam and Steger [4], Pulliam and Chaussee [5], and Pulliam [6].

The exercises at the end of the chapter provide an opportunity to write a computer program to apply this algorithm to several one-dimensional problems. Neither approximate factorization nor the coordinate transformation will enter into this program, but the exercise will enable the reader to develop a greater understanding of most other aspects of the algorithm.

4.1.1 Implicit Versus Explicit Time-Marching Methods

As discussed in Chap. 2, time-marching methods can be classified as implicit or explicit, and the two types have significantly different properties with respect to stability and cost. A simple characterization of implicit and explicit methods states that implicit methods have a much higher computing cost per time step, but their stability properties permit much larger time steps to be used. Depending on the nature of the problem, specifically its *stiffness*, either method can be more efficient. Implicit methods become relatively more efficient with increasing problem stiffness.

In computational fluid dynamics, stiffness has many sources, both physical and numerical. Physical stiffness comes from varying scales and speeds associated with different physical processes contained in the PDEs. For example, if the computation includes chemical reactions that proceed at rates much higher than those associated with the basic fluid dynamics, and time-accurate resolution of the chemical reactions is not required, then this will lead to a stiff system. Figure 2.2 shows one way in which numerical stiffness is introduced. There exist many modes in the system at high wavenumbers that are completely inaccurate. Such modes are inherently parasitic. This means that resolving them accurately in time will not improve the accuracy of the solution, because the spatial discretization is not accurate for these components of the solution. Thus these modes and their associated eigenvalues must lie within the stable region of the time-marching method, but need not lie within its region of accuracy (see Fig. 2.6). Furthermore, in many computations, very small grid spacings are needed in some regions of the flow, such as boundary layers, while much larger spacings are sufficient elsewhere. This too can cause stiffness, as the time taken for information to pass through a small cell is much shorter than that taken to pass through a large cell, introducing widely different time scales from a numerical point of view. Moreover, if gradients are much higher in one direction than another, then it is efficient to use small grid spacings in the direction of large gradient and larger spacings in the smaller gradient direction, leading to grid cells with high aspect ratios. As the time taken for waves to traverse the cell in one direction is thus much different from the other direction, multiple time scales and hence stiffness can again be introduced.

One way to understand the choice between implicit and explicit methods is to consider the limiting factor in the choice of the time step. Accuracy considerations place one bound on the maximum allowable time step. In other words, the time step must be small enough that the time accuracy of the solution is sufficient. Stability considerations place another bound on the time step. If the accuracy bound is smaller than the stability bound, then the time step is said to be *accuracy limited*. If the stability bound is smaller, then it is said to be *stability limited*. In a simulation where the time step is accuracy limited, there is little point in using an implicit method, as the same time step must be used in either case, so the extra cost per time step of an implicit method is not worthwhile. Conversely, if the stability bound is much smaller than the accuracy bound, then the explicit method will require a much smaller time step than an unconditionally stable implicit method, and hence the latter can be more efficient.

In the context of the numerical solution of ODEs, it is straightforward to categorize a method as explicit or implicit. In the context of PDEs, it is more accurate to classify methods according to a spectrum ranging from fully explicit to fully implicit. At the fully explicit end of the spectrum lies a method such as the explicit Euler method, without any additional convergence acceleration techniques, such as multigrid or implicit residual smoothing (which the reader will learn about in the next chapter). A multi-stage method, such as an explicit Runge-Kutta method, is still officially explicit, but generally has a larger stability bound at the expense of an increased cost per time step and can therefore be considered to have moved slightly toward the implicit end of the spectrum. Similarly, convergence acceleration techniques such as implicit residual smoothing and multigrid move the resulting "explicit" algorithm further in the implicit direction. This is typically associated with increased transfer of information across the mesh during a time step, which is a characteristic of implicit methods, an increased stability bound, and an increased cost per time step. At the fully implicit end of the spectrum lies the implicit Euler method with a direct solution of the linear problem at each time step. As this is usually infeasible and inefficient, for reasons to be discussed in this chapter, the linear problem is usually solved inexactly using an iterative method, which moves the algorithm slightly in the explicit direction. Alternatively, the linear problem can be approximated in a manner that makes it easier to solve, as in the approximate factorization algorithm that is the subject of this chapter. This reduces the cost per time step but can also reduce the optimal time step for convergence; in other words, it moves the algorithm somewhat further away from the fully implicit end of the spectrum.

Both the extreme explicit and the extreme implicit ends of the spectrum lead to inefficient algorithms for large problems. Therefore, all practical algorithms in use today for large-scale problems, including the algorithms described in this and the following chapter, lie somewhere between these two extremes, with the choice depending on the stiffness of the particular problem under consideration. It is interesting to note that, although this chapter's algorithm is nominally classified as implicit, while next chapter's algorithm is nominally classified as explicit, their cost per time step is quite comparable.

4.2 Generalized Curvilinear Coordinate Transformation

Finite-difference formulas are most naturally implemented on rectilinear meshes, as described in Chap. 2. On such meshes, the mesh lines are orthogonal, and it is straightforward to align the mesh such that each mesh line is associated with a specific coordinate direction. The derivative in a given coordinate direction can then be easily approximated based on finite differences along the corresponding mesh line. On the other hand, implementation of boundary conditions is simplified if the mesh is *body-fitted*, in other words the mesh conforms to the boundary of the geometry under consideration. If the boundary is curved, as is the case for most geometries of interest, this precludes the use of a mesh that is both rectilinear and body-fitted. In the present algorithm, this issue is addressed by transforming the physical space in which the mesh has curved, potentially non-orthogonal mesh lines into a computational space in which the mesh is rectilinear through a generalized curvilinear coordinate transformation. Such a transformation enables the straightforward application of finite-difference formulas on a body-fitted mesh. Our exposition will be in two dimensions, but extension to three dimensions should not present the reader with any conceptual difficulties.

An example of a mesh about an airfoil is shown in Fig. 4.1, and the corresponding curvilinear coordinate transformation is shown schematically in Fig. 4.2. In this case, the body is an airfoil, and the flow domain is bounded by an outer boundary. In the physical space defined by the Cartesian coordinates x, y, one set of mesh lines forms a "C" and hence such a mesh is known as a "C-mesh." The innermost "C" conforms to the airfoil surface and a *wake cut* along which two mesh lines correspond to a single line in physical space. The outermost "C" corresponds to the curved portion of the outer boundary. This set of lines is defined to be the one along which the curvilinear coordinate ξ varies, and the curvilinear coordinate η is constant. The second set of mesh lines is roughly orthogonal to the first and emanates from the body or the wake cut toward the outer boundary. Along these lines, η varies, and ξ is constant. The coordinate transformation is chosen such that the mesh is mapped to a computational space where the mesh lines are orthogonal, and the spacings $\Delta\xi$ and $\Delta\eta$ are unity in both directions. Therefore, standard finite-difference formulas can be easily applied. The computational space is a rectangle, where the bottom side includes the grid line lying on the airfoil and the wake cut, the top is the curved portion of the outer boundary, the left side is the portion of the back boundary below the wake cut, and the right side is the portion of the back boundary above the wake cut. Although

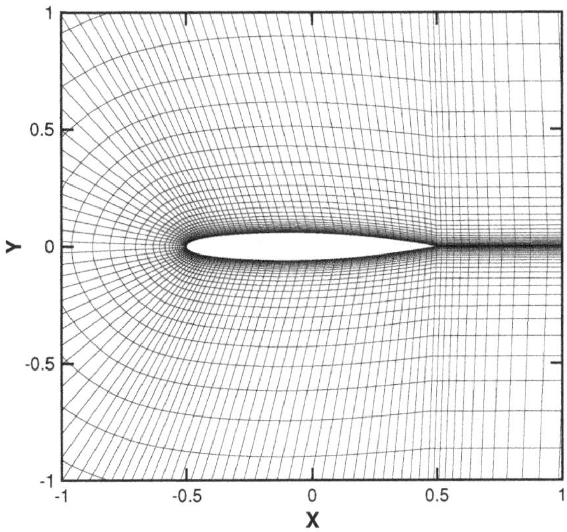

Fig. 4.1 A sample airfoil grid with a "C" topology showing only the region near the airfoil

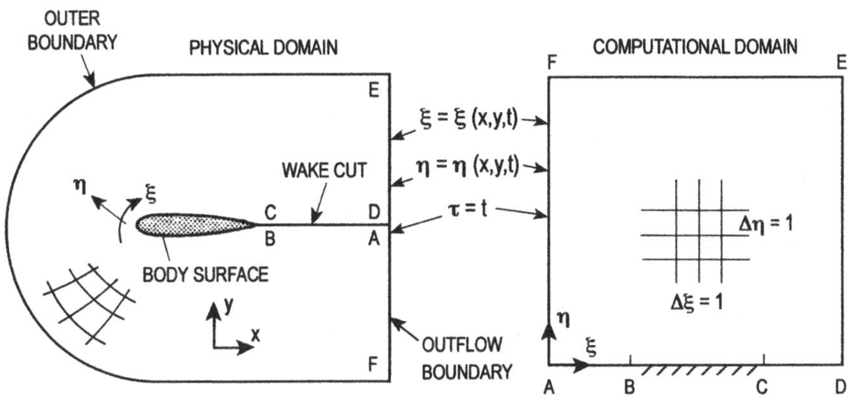

Fig. 4.2 An example of a generalized curvilinear coordinate transformation for a C-mesh

meshes can be defined by an analytical transformation for simple geometries, they are typically defined solely by the Cartesian coordinates of their nodes, and the underlying transformation to computational space is not known explicitly.

It is important to note that the mesh topology shown in Figs. 4.1 and 4.2 is just one possible topology. Another possibility, an "O" mesh, is shown in Fig. 4.3. The key property of such meshes, known as *structured meshes* is that the nodes are aligned along coordinate directions. This contrasts with *unstructured meshes*, which have no such constraint. An interior node in a two-dimensional structured mesh must have four neighbors (six in three dimensions), while a node in an unstructured mesh can have an arbitrary number of neighbors. This characteristic of a structured mesh sim-

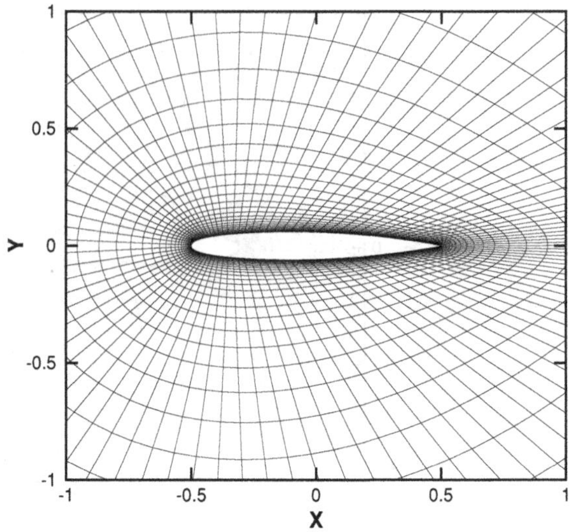

Fig. 4.3 A sample airfoil grid with an "O" topology showing only the region near the airfoil

plifies its storage. In two dimensions, a structured mesh is defined by a set of x and y coordinates that are assigned indices j and k, where j corresponds to the index in the ξ direction, and k corresponds to the η direction. The four immediate neighbors of node (j, k) are the nodes with indices $(j+1, k)$, $(j-1, k)$, $(j, k+1)$, $(j, k-1)$; the connectivity is implied by the indices. For more complex geometries, it can be impossible to define a mesh such that a single, simply connected, rectangular computational space exists. For such cases, block-structured meshes can be defined such that multiple rectangular computational domains are produced by the transformation. These domains can be interfaced in a number of different ways, including overlapping and abutting blocks.

In order to make use of finite-difference formulas defined in computational space, the governing equations must be transformed such that derivatives with respect to the Cartesian coordinates x and y are replaced by derivatives with respect to computational coordinates ξ and η. The coordinate transformation introduced here follows the development of Viviand [7] and Vinokur [8]. The Navier-Stokes equations can be transformed from Cartesian coordinates to generalized curvilinear coordinates where

$$\begin{aligned} \tau &= t \\ \xi &= \xi(x, y, t) \\ \eta &= \eta(x, y, t). \end{aligned} \tag{4.1}$$

If the grid does not deform over time, then $\xi = \xi(x, y)$ and $\eta = \eta(x, y)$. Typically there will be a one to one correspondence between a physical point in space and a computational point, except for regions where there are singularities or cuts due to the topology, such as the wake cut in the C-mesh example above. In those cases it may be necessary to map one physical point to more than one computational point.

The present coordinate transformation differs from some in that only the independent variables are transformed. The dependent variables remain defined in the Cartesian space, e.g. in terms of the Cartesian velocity components u and v. Chain-rule expansions are used to represent the derivatives in Cartesian space, ∂_t, ∂_x, and ∂_y of (3.1), in terms of the curvilinear derivatives, as follows:

$$\begin{aligned}
\frac{\partial}{\partial x} &= \frac{\partial \xi}{\partial x}\frac{\partial}{\partial \xi} + \frac{\partial \eta}{\partial x}\frac{\partial}{\partial \eta} \\
\frac{\partial}{\partial y} &= \frac{\partial \xi}{\partial y}\frac{\partial}{\partial \xi} + \frac{\partial \eta}{\partial y}\frac{\partial}{\partial \eta} \\
\frac{\partial}{\partial t} &= \frac{\partial}{\partial \tau} + \frac{\partial \xi}{\partial t}\frac{\partial}{\partial \xi} + \frac{\partial \eta}{\partial t}\frac{\partial}{\partial \eta}.
\end{aligned} \tag{4.2}$$

Introducing the notation

$$\partial_x \equiv \frac{\partial}{\partial x} \quad \text{and} \quad \xi_x \equiv \frac{\partial \xi}{\partial x}, \tag{4.3}$$

these can be written in matrix form as

$$\begin{bmatrix} \partial_t \\ \partial_x \\ \partial_y \end{bmatrix} = \begin{bmatrix} 1 & \xi_t & \eta_t \\ 0 & \xi_x & \eta_x \\ 0 & \xi_y & \eta_y \end{bmatrix} \begin{bmatrix} \partial_\tau \\ \partial_\xi \\ \partial_\eta \end{bmatrix}. \tag{4.4}$$

Applying these chain-rule expansions to the Navier-Stokes equations (3.1), we obtain

$$\begin{aligned}
&\partial_\tau Q + \xi_t \partial_\xi Q + \eta_t \partial_\eta Q + \xi_x \partial_\xi E + \eta_x \partial_\eta E + \xi_y \partial_\xi F + \eta_y \partial_\eta F \\
&\quad = Re^{-1}\left(\xi_x \partial_\xi E_\text{v} + \eta_x \partial_\eta E_\text{v} + \xi_y \partial_\xi F_\text{v} + \eta_y \partial_\eta F_\text{v}\right).
\end{aligned} \tag{4.5}$$

4.2.1 Metric Relations

In (4.5), derivatives with respect to t, x, and y have been replaced by derivatives with respect to τ, ξ, and η. Since the computational space is rectilinear and equally spaced, the latter can be easily approximated using finite-difference expressions—these will be presented in a subsequent section. The coefficients introduced ($\xi_t, \xi_x, \xi_y, \eta_t, \eta_x, \eta_y$) are known as grid metrics. Since in most cases the transformation from physical space to computational space is not known analytically, the metrics must be determined numerically. That is, we usually are provided with just the x, y coordinates of the grid points and must numerically generate the metrics $(\xi_t, \xi_x, \xi_y, \eta_t, \eta_x, \eta_y)$ using finite differences. This introduces a difficulty in that these are derivatives with respect to the original Cartesian coordinates.

In order to address this, consider the inverse of the transformation given in (4.1):

$$\begin{aligned} t &= \tau \\ x &= x(\xi, \eta, \tau) \\ y &= y(\xi, \eta, \tau). \end{aligned} \tag{4.6}$$

Reversing the role of the independent variables in the chain rule formulas (4.3), we have,

$$\partial_\tau = \partial_t + x_\tau \partial_x + y_\tau \partial_y, \quad \partial_\xi = x_\xi \partial_x + y_\xi \partial_y, \quad \partial_\eta = x_\eta \partial_x + y_\eta \partial_y, \tag{4.7}$$

which can be written in matrix form as

$$\begin{bmatrix} \partial_\tau \\ \partial_\xi \\ \partial_\eta \end{bmatrix} = \begin{bmatrix} 1 & x_\tau & y_\tau \\ 0 & x_\xi & y_\xi \\ 0 & x_\eta & y_\eta \end{bmatrix} \begin{bmatrix} \partial_t \\ \partial_x \\ \partial_y \end{bmatrix}. \tag{4.8}$$

Comparing (4.4) and (4.8), it is immediately clear that

$$\begin{bmatrix} 1 & \xi_t & \eta_t \\ 0 & \xi_x & \eta_x \\ 0 & \xi_y & \eta_y \end{bmatrix} = \begin{bmatrix} 1 & x_\tau & y_\tau \\ 0 & x_\xi & y_\xi \\ 0 & x_\eta & y_\eta \end{bmatrix}^{-1} \tag{4.9}$$

$$= J \begin{bmatrix} (x_\xi y_\eta - y_\xi x_\eta) & (-x_\tau y_\eta + y_\tau x_\eta) & (x_\tau y_\xi - y_\tau x_\xi) \\ 0 & y_\eta & -y_\xi \\ 0 & -x_\eta & x_\xi \end{bmatrix}, \tag{4.10}$$

where $J = (x_\xi y_\eta - x_\eta y_\xi)^{-1}$ is defined as the metric Jacobian. This yields the following metric relations:

$$\begin{aligned} \xi_t &= J(-x_\tau y_\eta + y_\tau x_\eta), \quad \xi_x = J y_\eta, \quad \xi_y = -J x_\eta \\ \eta_t &= J(x_\tau y_\xi - y_\tau x_\xi), \quad \eta_x = -J y_\xi, \quad \eta_y = J x_\xi. \end{aligned} \tag{4.11}$$

Using these relations, the metrics $(\xi_t, \xi_x, \xi_y, \eta_t, \eta_x, \eta_y)$ can be determined from $(x_\tau, x_\xi, x_\eta, y_\tau, y_\xi, y_\eta)$, where the latter are easily found using finite differences, since they are derivatives in computational space. Finite-difference formulas for these terms will be presented later in this chapter.

4.2.2 Invariants of the Transformation

At this point we notice that the transformed equations (4.5) are in a weak conservation law form. That is, even though none of the flow variables (or functions of the flow variables) occur as coefficients in the differential equations, the metrics, which are spatially varying, lie outside of the derivative operators. There is some argument in the literature which advocates the use of the so called "chain rule form," since it should still have good shock capturing properties and in some ways is a simpler form. Here, though, we shall restrict ourselves to the strong conservation law form which will be derived below.

To simplify our derivation, we will consider the inviscid terms only. This reduces (4.5) to

$$\partial_\tau Q + \xi_t \partial_\xi Q + \eta_t \partial_\eta Q + \xi_x \partial_\xi E + \eta_x \partial_\eta E + \xi_y \partial_\xi F + \eta_y \partial_\eta F = 0. \quad (4.12)$$

To produce the strong conservation law form we first multiply (4.12) by J^{-1} and apply the product rule to all terms. For example, the fourth term on the left-hand side can be expanded as

$$\left(\frac{\xi_x}{J}\right)\partial_\xi E = \partial_\xi\left(\frac{\xi_x}{J}E\right) - E\partial_\xi\left(\frac{\xi_x}{J}\right). \quad (4.13)$$

Each term can thus be rewritten as the difference between a term in the form we are looking for, with no coefficient outside the derivative operator, and a second term that is the product of a function of Q and a derivative of a quantity that is strictly a function of the grid. Collecting all the terms into two groups, with Term_1 representing the first group of terms and Term_2 the second, we obtain

$$\text{Term}_1 + \text{Term}_2 = 0,$$

where

$$\begin{aligned}
\text{Term}_1 &= \partial_\tau(Q/J) + \partial_\xi[(\xi_t Q + \xi_x E + \xi_y F)/J] + \partial_\eta[(\eta_t Q + \eta_x E + \eta_y F)/J] \\
\text{Term}_2 &= -Q[\partial_\tau(J^{-1}) + \partial_\xi(\xi_t/J) + \partial_\eta(\eta_t/J)] \qquad (4.14) \\
&\quad -E[\partial_\xi(\xi_x/J) + \partial_\eta(\eta_x/J)] - F[\partial_\xi(\xi_y/J) + \partial_\eta(\eta_y/J)].
\end{aligned}$$

The expressions from Term_2,

$$\begin{aligned}
&\partial_\tau(J^{-1}) + \partial_\xi(\xi_t/J) + \partial_\eta(\eta_t/J) \\
&\partial_\xi(\xi_x/J) + \partial_\eta(\eta_x/J) \\
&\partial_\xi(\xi_y/J) + \partial_\eta(\eta_y/J),
\end{aligned} \quad (4.15)$$

are defined as invariants of the transformation. Substituting the metric relations (4.11) into the invariant expressions gives

$$\partial_\tau(x_\xi y_\eta - y_\xi x_\eta) + \partial_\xi(-x_\tau y_\eta + y_\tau x_\eta) + \partial_\eta(x_\tau y_\xi - y_\tau x_\xi)$$

$$\partial_\xi(y_\eta) + \partial_\eta(-y_\xi) \tag{4.16}$$

$$\partial_\xi(-x_\eta) + \partial_\eta(x_\xi). \tag{4.17}$$

Analytically, differentiation is commutative, and the above terms sum to zero. This eliminates Term_2 of (4.15), and the resulting equations are in strong conservation law form.

There is an important issue associated with these invariants. It is not true in general that finite-difference approximations are commutative. Consequently, when numerical differencing is applied to these equations (as developed in the Sect. 4.4), the finite-difference formulas used to evaluate the spatial derivatives of the fluxes and the finite-difference formulas used to calculate the metrics do not necessarily satisfy the commutative law. Second-order central differences commute, but mixed second-order and fourth-order formulas do not. This is further discussed in Sect. 4.4.1.

4.2.3 Navier-Stokes Equations in Generalized Curvilinear Coordinates

The Navier-Stokes equations written in strong conservation law form are

$$\partial_\tau \widehat{Q} + \partial_\xi \widehat{E} + \partial_\eta \widehat{F} = Re^{-1}[\partial_\xi \widehat{E}_\text{v} + \partial_\eta \widehat{F}_\text{v}], \tag{4.18}$$

with

$$\widehat{Q} = J^{-1}\begin{bmatrix} \rho \\ \rho u \\ \rho v \\ e \end{bmatrix}, \widehat{E} = J^{-1}\begin{bmatrix} \rho U \\ \rho u U + \xi_x p \\ \rho v U + \xi_y p \\ U(e+p) - \xi_t p \end{bmatrix}, \widehat{F} = J^{-1}\begin{bmatrix} \rho V \\ \rho u V + \eta_x p \\ \rho v V + \eta_y p \\ V(e+p) - \eta_t p \end{bmatrix},$$

where

$$U = \xi_t + \xi_x u + \xi_y v, \quad V = \eta_t + \eta_x u + \eta_y v \tag{4.19}$$

are known as the contravariant velocity components—see Sect. 4.2.4 for more details. The viscous flux terms are $\widehat{E}_\text{v} = J^{-1}(\xi_x E_\text{v} + \xi_y F_\text{v})$ and $\widehat{F}_\text{v} = J^{-1}(\eta_x E_\text{v} + \eta_y F_\text{v})$.The viscous stress and heat conduction terms must also be transformed using the chain rule such that they are written in terms of ξ and η derivatives, giving

$$\begin{aligned}
\tau_{xx} &= \mu(4(\xi_x u_\xi + \eta_x u_\eta) - 2(\xi_y v_\xi + \eta_y v_\eta))/3 \\
\tau_{xy} &= \mu(\xi_y u_\xi + \eta_y u_\eta + \xi_x v_\xi + \eta_x v_\eta) \\
\tau_{yy} &= \mu(-2(\xi_x u_\xi + \eta_x u_\eta) + 4(\xi_y v_\xi + \eta_y v_\eta))/3 \\
f_4 &= u\tau_{xx} + v\tau_{xy} + \mu Pr^{-1}(\gamma - 1)^{-1}(\xi_x \partial_\xi a^2 + \eta_x \partial_\eta a^2) \\
g_4 &= u\tau_{xy} + v\tau_{yy} + \mu Pr^{-1}(\gamma - 1)^{-1}(\xi_y \partial_\xi a^2 + \eta_y \partial_\eta a^2).
\end{aligned} \tag{4.20}$$

The above discussion of metric invariants suggests a useful test for a finite-difference formulation. A minimum requirement of any finite-difference formulation is that a steady uniform flow be a valid solution of the discrete equations. If the chain-rule form (4.5) is evaluated for a steady uniform flow defined by

$$\begin{aligned}
\rho &= 1, \\
u &= M_\infty, \\
v &= 0, \\
e &= \frac{1}{\gamma(\gamma - 1)} + \frac{1}{2} M_\infty^2,
\end{aligned} \tag{4.21}$$

it is clearly satisfied, since all terms must equal zero given that the solution has no spatial or temporal variation. We would also like this steady uniform flow to satisfy (4.18) after the various derivatives have been replaced by finite-difference approximations. If the discrete form of (4.18) is not satisfied by a steady uniform flow, this can reveal a multitude of possible errors, including possibly a choice of difference operators for which the metric invariants are not zero.

4.2.4 Covariant and Contravariant Components in Curvilinear Coordinates

In Sect. 4.2.3 we introduced the contravariant velocity components associated with the curvilinear coordinate system. Since we will continue to work with Cartesian velocity components, a detailed knowledge of covariant and contravariant components is not necessary to understand the rest of the algorithm description. However, we will later need, for example, expressions for velocity components tangential and normal to a boundary in terms of the Cartesian components, so it is helpful to have a sufficient understanding to be able to derive such expressions.

We will assume a steady mesh in two dimensions, so we have $x(\xi, \eta)$, $y(\xi, \eta)$ and the inverse transformation $\xi(x, y)$, $\eta(x, y)$. First, define the vector

$$r = x\hat{i} + y\hat{j}. \tag{4.22}$$

In curvilinear coordinates, two sets of basis vectors can be defined. The covariant basis vectors are tangent to the ξ and η axes and are not required to be orthogonal.

They are given by

$$b_1 = \frac{\partial r}{\partial \xi}, \quad b_2 = \frac{\partial r}{\partial \eta}. \tag{4.23}$$

It can be more convenient to scale these such that they are unit vectors, giving

$$\hat{e}_1 = \frac{\frac{\partial r}{\partial \xi}}{\left|\frac{\partial r}{\partial \xi}\right|}, \quad \hat{e}_2 = \frac{\frac{\partial r}{\partial \eta}}{\left|\frac{\partial r}{\partial \eta}\right|}. \tag{4.24}$$

Note that these vectors are defined locally. The contravariant basis vectors are normal to the η and ξ axes and are defined by

$$B_1 = \nabla \xi, \quad B_2 = \nabla \eta, \tag{4.25}$$

where ∇ is the gradient operator. The contravariant basis vectors can also be scaled such that their length is unity:

$$\hat{E}_1 = \frac{\nabla \xi}{|\nabla \xi|}, \quad \hat{E}_2 = \frac{\nabla \eta}{|\nabla \eta|}. \tag{4.26}$$

With these bases, an arbitrary vector A can be defined in the following ways:

$$\begin{aligned} A &= A_1 \hat{e}_1 + A_2 \hat{e}_2 = a_1 \hat{E}_1 + a_2 \hat{E}_2 \\ &= C_1 b_1 + C_2 b_2 = c_1 B_1 + c_2 B_2. \end{aligned} \tag{4.27}$$

Here C_1 and C_2 are the contravariant components of A, i.e. $C_1 = B_1 \cdot A$ and $C_2 = B_2 \cdot A$, and c_1 and c_2 are the covariant components of A, i.e. $c_1 = b_1 \cdot A$ and $c_2 = b_2 \cdot A$. Note that $B_i \cdot b_j = \delta_{ij}$, where δ_{ij} is the Kronecker delta.

For example, let A represent the velocity vector $u\hat{i} + v\hat{j}$. From (4.25) we have

$$B_1 = \xi_x \hat{i} + \xi_y \hat{j}, \quad B_2 = \eta_x \hat{i} + \eta_y \hat{j}. \tag{4.28}$$

Therefore, we obtain for the contravariant components of velocity

$$C_1 = B_1 \cdot A = \xi_x u + \xi_y v, \quad C_2 = B_2 \cdot A = \eta_x u + \eta_y v, \tag{4.29}$$

consistent with the definitions of U and V in (4.19) when the coordinate transformation is time invariant.

In the application of boundary conditions, one often needs expressions for the velocity components normal and tangential to the boundary in terms of the Cartesian velocity components. In this case, we must work with unit basis vectors to preserve the magnitude of the velocity. We assume that the boundary is a grid line of constant η, such as the airfoil surface in Figs. 4.1 and 4.2, but the result is easily generalized to other boundaries. Recall that $\hat{e}_1$ is tangent to the ξ axis, and $\hat{E}_2$ is normal to the η axis. Therefore, we can write

$$u\hat{i} + v\hat{j} = V_t\hat{e}_1 + V_n\hat{E}_2, \tag{4.30}$$

where V_t and V_n are the tangential and normal velocity components, respectively. The two unit vectors are given by

$$\hat{e}_1 = \frac{x_\xi\hat{i} + y_\xi\hat{j}}{\sqrt{x_\xi^2 + y_\xi^2}} = \frac{\eta_y\hat{i} - \eta_x\hat{j}}{\sqrt{\eta_x^2 + \eta_y^2}}$$
$$\hat{E}_2 = \frac{\eta_x\hat{i} + \eta_y\hat{j}}{\sqrt{\eta_x^2 + \eta_y^2}}, \tag{4.31}$$

where the metric relations are used to obtain the second expression for $\hat{e}_1$. Noting that

$$\hat{e}_1 \cdot \hat{E}_2 = 0, \quad \hat{e}_1 \cdot \hat{e}_1 = \hat{E}_2 \cdot \hat{E}_2 = 1, \tag{4.32}$$

we find the following expressions for the tangential and normal velocity components:

$$V_t = \hat{e}_1 \cdot (u\hat{i} + v\hat{j}) = \frac{\eta_y u - \eta_x v}{\sqrt{\eta_x^2 + \eta_y^2}}$$
$$V_n = \hat{E}_2 \cdot (u\hat{i} + v\hat{j}) = \frac{\eta_x u + \eta_y v}{\sqrt{\eta_x^2 + \eta_y^2}}. \tag{4.33}$$

These are the velocity components tangential and normal to a grid line of constant η at a specific point in space.

As a second example, consider the derivative of pressure in a direction normal to a surface which again corresponds to a grid line of constant η. The gradient of pressure can be expressed in terms of the basis vectors $\hat{e}_1$ and $\hat{E}_2$ as follows:

$$\nabla p = \frac{\partial p}{\partial x}\hat{i} + \frac{\partial p}{\partial y}\hat{j} = \frac{\partial p}{\partial t}\hat{e}_1 + \frac{\partial p}{\partial n}\hat{E}_2, \tag{4.34}$$

where here t refers to the tangential coordinate. The normal derivative can be isolated by taking the dot product with $\hat{E}_2$ (which is identical to $\hat{n}$):

$$\frac{\partial p}{\partial n} = \hat{E}_2 \cdot \nabla p = \frac{\eta_x \frac{\partial p}{\partial x} + \eta_y \frac{\partial p}{\partial y}}{\sqrt{\eta_x^2 + \eta_y^2}}. \tag{4.35}$$

The chain rule gives

$$\frac{\partial p}{\partial x} = \eta_x \frac{\partial p}{\partial \eta} + \xi_x \frac{\partial p}{\partial \xi}, \qquad \frac{\partial p}{\partial y} = \eta_y \frac{\partial p}{\partial \eta} + \xi_y \frac{\partial p}{\partial \xi}, \tag{4.36}$$

from which we obtain the final expression for the normal derivative:

$$\frac{\partial p}{\partial n} = \frac{(\eta_x \xi_x + \eta_y \xi_y)\frac{\partial p}{\partial \xi} + (\eta_x^2 + \eta_y^2)\frac{\partial p}{\partial \eta}}{\sqrt{\eta_x^2 + \eta_y^2}}. \tag{4.37}$$

4.3 Thin-Layer Approximation

We introduce the thin-layer approximation [9] here only to simplify the treatment of the viscous terms in the exposition of the algorithm. It is not of fundamental importance and is applicable only if the following criteria are satisfied:

- The Reynolds number is high; the geometry is streamlined and at a modest angle of incidence with respect to the flow direction. Consequently, boundary layers remain attached or mildly separated, and both boundary layers and wakes are thin relative to the characteristic dimension of the geometry.
- The mesh is body fitted, and mesh lines are at least close to orthogonal to the surface, as depicted in Fig. 4.4. Moreover, lines of constant η are reasonably well aligned with wakes. As a result of this last constraint, a C-mesh is a better choice than an O-mesh when the thin-layer approximation is used.

Under these conditions, boundary-layer theory shows that streamwise gradients of viscous and turbulent stresses are small compared to normal gradients in boundary layers and wakes, and viscous and turbulent stresses are negligible outside of boundary layers and wakes. Therefore, mesh resolution requirements typically dictate a smaller mesh spacing in the direction normal to the surface in boundary layers, leading to meshes with cells having high aspect ratios near the surface, as in Fig. 4.4. Moreover, streamwise gradients of viscous and turbulent stresses can often be neglected with little impact on solution accuracy, leading to the thin-layer Navier-Stokes equations. It is important to recognize that although the rationale for the thin-layer Navier-Stokes equations is closely related to that for the boundary-layer equations, unlike the latter, the thin-layer Navier-Stokes equations retain all inviscid terms in full. Hence they are applicable both within boundary layers and wakes and outside these regions, where the flow is effectively inviscid.

We will assume that mesh lines along which η varies are nearly normal to the surface, as shown in Fig. 4.4. Applying the thin-layer approximation to (4.18) then involves neglecting the term $\partial_\xi \widehat{E}_v$ as well as all derivatives with respect to ξ in $\widehat{F}_v$, leading to

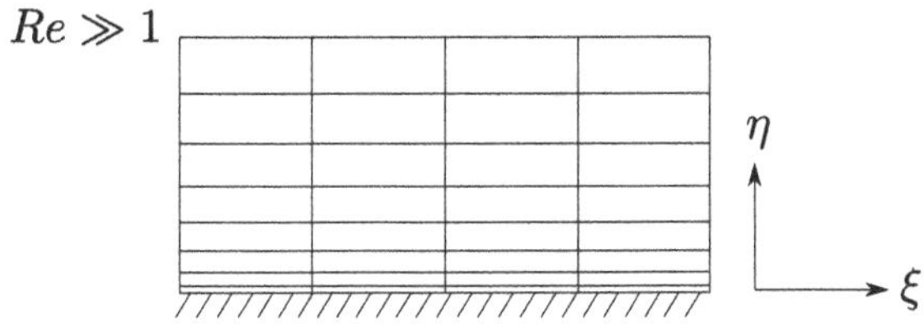

Fig. 4.4 Mesh near body surface

$$\partial_\tau \widehat{Q} + \partial_\xi \widehat{E} + \partial_\eta \widehat{F} = Re^{-1} \partial_\eta \widehat{S}, \tag{4.38}$$

where

$$\widehat{S} = J^{-1} \begin{bmatrix} 0 \\ \eta_x m_1 + \eta_y m_2 \\ \eta_x m_2 + \eta_y m_3 \\ \eta_x (u m_1 + v m_2 + m_4) + \eta_y (u m_2 + v m_3 + m_5) \end{bmatrix}, \tag{4.39}$$

with

$$\begin{aligned} m_1 &= \mu(4\eta_x u_\eta - 2\eta_y v_\eta)/3 \\ m_2 &= \mu(\eta_y u_\eta + \eta_x v_\eta) \\ m_3 &= \mu(-2\eta_x u_\eta + 4\eta_y v_\eta)/3 \\ m_4 &= \mu Pr^{-1}(\gamma - 1)^{-1} \eta_x \partial_\eta (a^2) \\ m_5 &= \mu Pr^{-1}(\gamma - 1)^{-1} \eta_y \partial_\eta (a^2). \end{aligned} \tag{4.40}$$

Although the thin-layer approximation was quite popular in the early days of CFD, it is important for the reader to understand that the algorithms presented here do not depend on this approximation and are applicable to the full Navier-Stokes equations. We proceed with the thin-layer approximation only because it simplifies our presentation of the algorithms while retaining their key features.

4.4 Spatial Differencing

We will now present an algorithm for the numerical solution of the transformed Navier Stokes equations (4.18), which in turn will provide a solution to the original equations in Cartesian coordinates (3.1). The algorithm will follow the semi-discrete approach described in Chap. 2 in which the spatial derivatives are approximated first to produce a system of ODEs.

Whether we are interested in a steady solution or a time-accurate solution to an unsteady problem, the first step is to take the continuous differential operators ∂_ξ and ∂_η and approximate them with finite-difference operators on a discrete mesh. This is facilitated by the use of the generalized curvilinear coordinate transformation described in Sect. 4.2. A structured mesh is defined by a set of coordinate pairs $x(j,k)$, $y(j,k)$, where j and k are integer indices. If one defines $\xi \equiv j$ and $\eta \equiv k$, then the grid spacing in the computational space is unity in both directions, that is

$$\Delta\xi = 1, \quad \Delta\eta = 1. \tag{4.41}$$

Since the mesh is rectilinear and uniform in computational space, one can apply finite-difference formulas in a straightforward manner. We will use subscripts to indicate the coordinates of a flow variable in computational space, i.e.

$$Q_{j,k} := Q(j\Delta\xi, k\Delta\eta). \tag{4.42}$$

We can use second-order centered difference operators for the inviscid flux derivatives $\partial_\xi \widehat{E}$ and $\partial_\eta \widehat{F}$ as follows:

$$\delta_\xi \widehat{E}_{j,k} = \frac{\widehat{E}_{j+1,k} - \widehat{E}_{j-1,k}}{2\Delta\xi}, \quad \delta_\eta \widehat{F}_{j,k} = \frac{\widehat{F}_{j,k+1} - \widehat{F}_{j,k-1}}{2\Delta\eta}. \tag{4.43}$$

Similarly, second-order centered differences can be used for the metric terms, such as

$$\left(x_\xi\right)_{j,k} = \frac{x_{j+1,k} - x_{j-1,k}}{2\Delta\xi}. \tag{4.44}$$

Since $\Delta\xi = \Delta\eta = 1$ as a result of the transformation to computational space, we omit these terms for the remainder of this presentation.

For the viscous derivatives, the terms take the form

$$\partial_\eta \left(\alpha_{j,k}\partial_\eta \beta_{j,k}\right), \tag{4.45}$$

where $\alpha_{j,k}$ represents a spatially varying coefficient, such as a grid metric or the fluid viscosity, and $\beta_{j,k}$ is a velocity component or the square of the sound speed. Such a term can be approximated by differencing $\partial_\eta \beta_{j,k}$ using a second-order centered difference at each node, multiplying by the spatially varying coefficient, and applying the centered first-derivative approximation again. However, this leads to a five-point stencil involving values from $k-2$ to $k+2$ in the evaluation of (4.45). In the interest of retaining a compact three-point form, the term $\partial_\eta \beta_{j,k}$ can instead be evaluated at intermediate locations $k - \frac{1}{2}$ and $k + \frac{1}{2}$ using the following centered difference formulas:

$$\left(\frac{\partial \beta}{\partial \eta}\right)_{k+1/2} = \beta_{j,k+1} - \beta_{j,k}$$
$$\left(\frac{\partial \beta}{\partial \eta}\right)_{k-1/2} = \beta_{j,k} - \beta_{j,k-1}. \tag{4.46}$$

To second-order accuracy, the values of the spatially varying coefficient at the intermediate nodes can be found by averaging as follows:

$$\alpha_{j,k+1/2} = \frac{1}{2}\left(\alpha_{j,k} + \alpha_{j,k+1}\right)$$
$$\alpha_{j,k-1/2} = \frac{1}{2}\left(\alpha_{j,k-1} + \alpha_{j,k}\right). \tag{4.47}$$

A compact three-point finite-difference approximation to (4.45) can be obtained by applying a centered difference approximation at j, k using the intermediate points $j, k + \frac{1}{2}$ and $j, k - \frac{1}{2}$, as follows:

$$\frac{\left(\alpha_{j,k+1} + \alpha_{j,k}\right)}{2}\left(\beta_{j,k+1} - \beta_{j,k}\right) - \frac{\left(\alpha_{j,k} + \alpha_{j,k-1}\right)}{2}\left(\beta_{j,k} - \beta_{j,k-1}\right). \tag{4.48}$$

We will consider only second-order schemes in this chapter, but higher-order operators, as described in Sect. 2.2, can offer improved efficiency in certain contexts. If higher-order differencing operators are used, the metric terms should also be evaluated using the same first-derivative operator, boundary schemes of appropriate order[1] should be used, and the accuracy of other approximations in the algorithm, such as numerical integration to obtain forces, should also be raised to a consistent order.

At this point it is reasonable to ask whether the second-order centered difference formula remains second order on a nonuniform mesh when a curvilinear coordinate transformation is used. To address this question, consider a nonuniform mesh in one dimension, for which the coordinate transformation gives for a first derivative

$$\frac{\partial f}{\partial x} = \xi_x \frac{\partial f}{\partial \xi} = \frac{1}{x_\xi}\frac{\partial f}{\partial \xi}. \tag{4.49}$$

Application of second-order centered difference formulas to both $\partial f/\partial \xi$ and x_ξ at node j gives

$$(\delta_x f)_j = \frac{f_{j+1} - f_{j-1}}{x_{j+1} - x_{j-1}}. \tag{4.50}$$

[1] The order of a boundary scheme can often be one less than that of the interior scheme, and the global accuracy of the interior operator is still achieved (see Gustafsson [10]). The stability of boundary schemes for higher-order methods is also an important consideration but is beyond the scope of this book.

Denoting the mesh spacing immediately to the right of node j as

$$\Delta x_+ = x_{j+1} - x_j, \tag{4.51}$$

and that to the left as

$$\Delta x_- = x_j - x_{j-1}, \tag{4.52}$$

a Taylor series expansion of the derivative operator gives the following error term:

$$\frac{1}{2}\left(\frac{\partial^2 f}{\partial x^2}\right)_j (\Delta x_+ - \Delta x_-) + \frac{1}{6}\left(\frac{\partial^3 f}{\partial x^3}\right)_j \left(\frac{\Delta x_+^3 + \Delta x_-^3}{\Delta x_+ + \Delta x_-}\right) + \cdots . \tag{4.53}$$

The second term is clearly second order, but, at first glance, the first term appears to be first order. However, it is important to recall that the notion of the order of accuracy relates to the behavior of the error when a smooth mesh is refined uniformly.

For our present example, we can define a mesh function $x(\xi) = g(\xi/M) = g(\xi D)$, where M is the number of cells in the one-dimensional mesh, and $D = 1/M$ is a nominal mesh spacing parameter. For example, if the number of nodes M is doubled, then D is halved. With this mesh function, Taylor series expansions for Δx_+ and Δx_- give

$$\Delta x_+ = x_{j+1} - x_j = Dg'_j + \frac{1}{2}D^2 g''_j + \frac{1}{6}d^3 g'''_j + \cdots \tag{4.54}$$

and

$$\Delta x_- = x_j - x_{j-1} = Dg'_j - \frac{1}{2}D^2 g''_j + \frac{1}{6}d^3 g'''_j - \cdots . \tag{4.55}$$

Taking the difference gives

$$\Delta x_+ - \Delta x_- = D^2 g''_j + \cdots = O(D^2), \tag{4.56}$$

and we see that the error term remains second order, even on a nonuniform mesh. It is important to note that the error (4.53) contains a term proportional to $\partial^2 f/\partial x^2$, so, although it is second order, this approximation is not exact for a quadratic function, as is the case on a uniform mesh, where $\Delta x_+ - \Delta x_-$ is zero. One can easily define a finite-difference scheme on a nonuniform mesh that is exact for a quadratic function, but this approach extends to multiple dimensions in a straightforward manner only if the mesh is rectangular.

In order to make the above discussion more concrete, consider the one-dimensional mesh function

$$x(\xi) = \frac{e^{\xi/M} - 1}{e - 1}. \tag{4.57}$$

This function produces a uniform stretching ratio given by

$$\frac{\Delta x_+}{\Delta x_-} = \frac{e^{1/M} - 1}{1 - e^{-1/M}}. \tag{4.58}$$

With $M = 10$, the stretching ratio is roughly 1.105; if M is increased to 100, the stretching ratio is reduced to roughly 1.010. With each increase in M, not only does the mesh spacing decrease in proportion to $1/M$, but the stretching ratio also decreases. Consequently, the difference $\Delta x_+ - \Delta x_-$ is of order $(1/M)^2$. If mesh refinement is performed such that the stretching ratio is constant, this is not a suitable refinement and the second-order behaviour of the difference operator (4.50) will not be observed .

4.4.1 Metric Differencing and Invariants

The second-order centered difference formulas used in two dimensions naturally produce consistent metric invariants, but in three dimensions some additional measures must be taken to ensure this property. Examining one of these terms in two dimensions, $\partial_\xi(y_\eta) + \partial_\eta(-y_\xi)$, using second-order centered differences both to form the metric terms and to approximate the flux derivatives, we obtain

$$\begin{aligned} \delta_\xi\delta_\eta y_{j,k} - \delta_\eta\delta_\xi y_{j,k} &= \delta_\xi(y_{j,k+1} - y_{j,k-1})/2 - \delta_\eta(y_{j+1,k} - y_{j-1,k})/2 \\ &= [y_{j+1,k+1} - y_{j-1,k+1} - y_{j+1,k-1} + y_{j-1,k-1}]/4 \\ &\quad -[y_{j+1,k+1} - y_{j+1,k-1} - y_{j-1,k+1} + y_{j-1,k-1}]/4 \\ &= 0, \end{aligned} \tag{4.59}$$

as desired.

In three dimensions, there are several different ways to ensure that these terms are zero. For example, consider the metric ξ_x, which is given by[2]

$$\xi_x = J(y_\eta z_\zeta - y_\zeta z_\eta), \tag{4.60}$$

where z and ζ are the third coordinate directions in Cartesian and computational space, respectively. One approach is to form ξ_x through the following formula:

$$\xi_x = J\left[(\mu_\zeta\delta_\eta y)(\mu_\eta\delta_\zeta z) - (\mu_\eta\delta_\zeta y)(\mu_\zeta\delta_\eta z)\right], \tag{4.61}$$

where δ is the second-order centered difference operator, and μ is an averaging operator defined, for example, by $\mu_\eta x_{j,k,l} = (x_{j,k+1,l} + x_{j,k-1,l})/2$, where l is the

[2] See Sect. 4.7.1

index in the ζ direction. If all of the metric terms are calculated in this manner, then the metric invariants will be satisfied.

An alternative approach in three dimensions that extends to higher order involves writing the expression for ξ_x as [11]:

$$\xi_x = J((y_\eta z)_\zeta - (y_\zeta z)_\eta), \tag{4.62}$$

which is analytically equivalent to (4.60). Analogous expressions can be written for the other metrics of the transformation. If consistent centered difference formulas are used for both the derivatives in such expressions for the metric terms as well as the flux derivatives, e.g. $\delta_\xi \widehat{E}$, then the metric invariants will be zero (within the limits of round-off error).

In (4.59) we saw that second-order centered differencing of both the metric relations and the flux derivatives leads to satisfaction of the invariant relations in two dimensions. However, consider the case of centered differencing to form the metrics combined with first-order one-sided backward differencing for the fluxes. We obtain

$$\begin{aligned}\nabla_\xi \delta_\eta y - \nabla_\eta \delta_\xi y = {} & [y_{j,k+1} - y_{j-1,k+1} - y_{j,k-1} + y_{j-1,k-1}]/2 \\ & + [-y_{j+1,k} + y_{j+1,k-1} + y_{j-1,k} - y_{j-1,k-1}]/2 \neq 0.\end{aligned} \tag{4.63}$$

The error associated with not satisfying the invariant relations is a truncation error that corresponds to the order of the lowest-order-accurate operator used or higher.

4.4.2 Artificial Dissipation

The concept of numerical dissipation was introduced in Sect. 2.5. Numerical dissipation can be added to a spatial discretization for three distinct purposes:

- to eliminate high-frequency modes that are not resolved and can contaminate the solution;
- to enhance stability and convergence to steady state;
- to prevent oscillations at discontinuities, such as shock waves.

The idea is to achieve these three purposes by introducing a level of numerical dissipation that does not significantly increase the overall numerical error.

In linear problems, such as the linear convection equation, the frequencies or wavenumbers present in the solution are dictated by the initial and boundary conditions. In the numerical solution of such equations, the components with wavenumbers that are not well resolved (see Fig. 2.2) are essentially spurious. They will not be handled accurately by the numerical scheme in terms of either convection or diffusion. Therefore, it can be worthwhile to remove them through numerical dissipation or filtering.

In solutions of the Euler and Navier-Stokes equations, nonlinear interactions occur between waves as a result of the nonlinearity in the convection terms of the momentum equations. If scale is represented by wavelength or frequency, it can be shown that two waves interact as products to form a wave of higher frequency (the sum of the original two) and one of lower frequency (the difference). In a physical system, this can lead to turbulence and the formation of shock waves. As a result of viscosity, there is a limit to the smallest length scales that arise. Numerically, if all scales are well resolved, for example in a well-resolved direct numerical simulation of a turbulent flow or a well-resolved simulation of a laminar flow, then numerical dissipation is not needed. However, in most flow computations, these smallest scales are typically not resolved. As a result, the true physical mechanism that puts an upper bound on the frequencies present in the solution is not accurately represented in the numerical solution. The lower frequencies do not cause a problem, but the continual cascading into higher and higher frequencies can lead to instabilities. These can be addressed through numerical dissipation. Even in linear problems, instabilities can arise from numerical implementation of boundary conditions and other approximations that might cause some eigenvalues of the semi-discrete operator matrix to lie slightly in the right half-plane. Numerical dissipation can address such instabilities as well and speed up convergence to a steady state.

The Euler equations support discontinuities such as shock waves, slip lines, and contact surfaces. Across these discontinuities, the differential form of the PDEs does not apply, so the appropriate jump conditions must be determined from the integral form. In essence, shock waves are a limiting case of the frequency cascade described in the previous paragraph. The Euler equations contain no mechanism to limit the minimum length scale, so a shock wave is a true discontinuity in an inviscid flow. In a real viscous flow, shock waves have a finite thickness, but it is so small that it is rarely practical to resolve a shock, and in any case it is not clear that the continuum hypothesis would be applicable within a shock wave. Therefore, although the Navier-Stokes equations do not support discontinuities, the issue of the numerical treatment of shock waves is present even in computations of viscous flows. Without a careful treatment, oscillations will occur at and near shock waves and other discontinuities.

Historically, the numerical treatment of shock waves has been divided into two approaches, shock fitting and shock capturing. In shock fitting, one must know the location of the shock and apply the jump conditions across it. While this is an inherently elegant approach, in practice it is very difficult to track the precise location of shock waves. As a result, shock capturing, in which the shock wave is smoothed out by numerical dissipation and the flow is treated as if it were continuous, has become the predominant approach.

A substantial amount of research has gone into the development of numerical methods for capturing shocks. We will cover such methods in more detail in Chap. 6. For our purpose here it suffices to say that in order to prevent oscillations, first-order numerical dissipation is needed in the vicinity of discontinuities. However, use of first-order numerical dissipation throughout the flow domain would lead to very large numerical errors, or, alternatively, the need for a very fine mesh to reduce numerical errors to the desired levels. Consequently, the numerical dissipation added to a spatial discretization of the Euler or Navier-Stokes equations generally consists of the following three components:

- a high-order component for smooth regions of the flow field,
- a first-order component for shock capturing,
- a means of sensing shocks and other discontinuities so that the appropriate dissipation operator can be selected in different regions of the flow field.

Before continuing, the reader may wish to review Sect. 2.5, which introduced the basic concepts underlying numerical dissipation. The dissipation is associated with the symmetric part of the difference operator and can be added either explicitly through artificial dissipation or through one-sided or upwind schemes that inherently include a symmetric component. In this chapter and the next we will concentrate on centered schemes with added artificial dissipation, while Chap. 6 discusses upwind schemes in more detail. The close relationship between the two approaches is clear from Sect. 2.5.

4.4.3 A Nonlinear Artificial Dissipation Scheme

Recalling Sect. 2.5, numerical dissipation can be added to a centered differencing scheme by adding a symmetric component to the difference operator approximating the first derivatives in the inviscid flux terms. For a constant-coefficient, linear hyperbolic system of equations in the form

$$\frac{\partial u}{\partial t} + \frac{\partial f}{\partial x} = \frac{\partial u}{\partial t} + A\frac{\partial u}{\partial x} = 0, \tag{4.64}$$

where $f = Au$, the dissipation can be added in the following manner:

$$\delta_x f = \delta_x^{\mathrm{a}} f + \delta_x^{\mathrm{s}}(|A|u), \tag{4.65}$$

where δ_x^{a} and δ_x^{s} are antisymmetric and symmetric difference operators, X is the matrix of right eigenvectors of A, Λ is a diagonal matrix containing the eigenvalues of A, and $|A| = X|\Lambda|X^{-1}$. The antisymmetric operator is simply the centered difference scheme, and the symmetric operator introduces the dissipation.

An antisymmetric or centered difference operator for a first derivative has an even order of accuracy, while the symmetric term has an odd order of accuracy. For smooth regions of the flow, the symmetric operator should be at least third order, since a first-order term is generally too dissipative and will add too much numerical error. With second-order centered differences, a third-order dissipation term is thus a good choice for regions where the flow variables behave smoothly, i.e. away from discontinuities.

Consequently, the following symmetric operator is often used together with second-order centered differences:

$$\left(\delta_x^s u\right)_j = \frac{\epsilon_4}{\Delta x}(u_{j-2} - 4u_{j-1} + 6u_j - 4u_{j+1} + u_{j+2}) \propto \epsilon_4 \Delta x^3 \frac{\partial^4 u}{\partial x^4}, \quad (4.66)$$

where ϵ_4 is a user defined constant. This operator is sufficient to damp unwanted high-frequency modes and provide stability while generally adding an error that is smaller than the second-order error associated with the centered difference scheme. However, it is not sufficient to prevent oscillations at discontinuities. For this purpose, the following first-order symmetric operator is typically used:

$$\left(\delta_x^s u\right)_j = \frac{\epsilon_2}{\Delta x}(-u_{j-1} + 2u_j - u_{j+1}) \propto -\epsilon_2 \Delta x \frac{\partial^2 u}{\partial x^2}. \quad (4.67)$$

The artificial dissipation scheme used in the implicit finite-difference algorithm of this chapter combines the above two operators using a pressure sensor to detect shock waves [6, 12]. This approach is intended for flows with shock waves, where the pressure is discontinuous; it will not sense a discontinuity such as a contact surface across which the pressure is continuous. Before presenting the operator, we note that

$$\nabla\Delta\nabla\Delta u_j = u_{j-2} - 4u_{j-1} + 6u_j - 4u_{j+1} + u_{j+2} \quad (4.68)$$

and

$$\nabla\Delta u_j = u_{j-1} - 2u_j + u_{j+1}, \quad (4.69)$$

where $\nabla u_j = u_j - u_{j-1}$ and $\Delta u_j = u_{j+1} - u_j$ are undivided differences.

Before moving to the two-dimensional equations in curvilinear coordinates, let us first consider the one-dimensional Euler equations (3.24):

$$\frac{\partial Q}{\partial t} + \frac{\partial E}{\partial x} = 0, \quad (4.70)$$

where $E = AQ$ as a result of the homogeneous property of the Euler equations (see [13], Appendix C). A natural application of (4.65) and (4.66) gives a fourth-difference dissipative term in the following form:

$$D_j = \nabla\Delta\nabla\Delta|A_j|Q_j. \quad (4.71)$$

In the constant-coefficient, linear case, $|A|$ is constant, but that is no longer true in the nonlinear case, and hence its position in the above equation can have a significant effect. For example, the choice

$$D_j = |A_j|\nabla\Delta\nabla\Delta Q_j \quad (4.72)$$

is not conservative. The preferred choice, motivated by analogy to flux-difference splitting, is

$$D_j = \nabla|A_{j+1/2}|\Delta\nabla\Delta Q_j, \tag{4.73}$$

where $A_{j+1/2}$ is some sort of average, such as a simple average or a Roe average (see Sect. 6.3).

Now consider the strong conservation law form of the Navier-Stokes equations in generalized curvilinear coordinates (4.18) with the spatial derivatives replaced by second-order centered differences, as in (4.43), and all of the spatial terms moved to the right-hand side:

$$\partial_\tau \widehat{Q} = -\delta_\xi \widehat{E} - \delta_\eta \widehat{F} + Re^{-1}[\delta_\xi \widehat{E}_v + \delta_\eta \widehat{F}_v], \tag{4.74}$$

where the compact three-point form (4.48) is assumed for the viscous derivatives. Let us restrict our interest for now to the inviscid term in the ξ direction, giving

$$\partial_\tau \widehat{Q} = -\delta_\xi \widehat{E}. \tag{4.75}$$

This can be written in the *conservation form*

$$\partial_\tau \widehat{Q} = -(f_{j+1/2} - f_{j-1/2}), \tag{4.76}$$

where

$$f_{j+1/2} = \frac{1}{2}(\widehat{E}_j + \widehat{E}_{j+1}). \tag{4.77}$$

Thus the discrete form applied to the conservation law form of the equation preserves the conservative property of the original PDE. It is important that the artificial dissipation scheme maintain this property.

We now introduce an artificial dissipation term $(D_\xi)_{j,k}$ in the ξ direction into (4.75) as follows:

$$(\partial_\tau \widehat{Q})_{j,k} = -(\delta_\xi \widehat{E})_{j,k} + (D_\xi)_{j,k}, \tag{4.78}$$

where

$$\begin{aligned}(D_\xi)_{j,k} = {} & \nabla_\xi \left(\epsilon^{(2)}|\widehat{A}|J^{-1}\right)_{j+1/2,k} \Delta_\xi Q_{j,k} \\ & -\nabla_\xi \left(\epsilon^{(4)}|\widehat{A}|J^{-1}\right)_{j+1/2,k} \Delta_\xi \nabla_\xi \Delta_\xi Q_{j,k}.\end{aligned} \tag{4.79}$$

Analogous terms are used in the η direction. There are many aspects to this expression; these will be explained one at a time. The first term on the right-hand side is the second-difference term, which is first order, that is needed near shock waves. The second term is the fourth-difference term, which is third order, that is used in smooth regions of the flow field. Their relative contributions are controlled by the two coefficients $\epsilon^{(2)}$ and $\epsilon^{(4)}$, which are defined below. Next, $\widehat{A}$ is the flux Jacobian in the ξ direction defined as follows:

$$\widehat{A} = \frac{\partial \widehat{E}}{\partial \widehat{Q}}. \tag{4.80}$$

This is given in Sect. 4.5.

Notice that the dissipation operates on Q, not $\widehat{Q}$; J^{-1} is moved together with $|\widehat{A}|$. This ensures that no dissipation is generated for a uniform flow. On a nonuniform mesh, $\widehat{Q}$ is not constant in space, even if Q is constant, as a result of the spatial variation of J^{-1}. Consequently, nonzero dissipation would arise in a uniform flow if the dissipation were to operate on $\widehat{Q}$.

The location of the terms $\epsilon^{(2)}|\widehat{A}|J^{-1}$ and $\epsilon^{(4)}|\widehat{A}|J^{-1}$ is consistent with (4.73). These can be evaluated through simple averages, e.g.

$$\left(\epsilon^{(2)}|\widehat{A}|J^{-1}\right)_{j+1/2,k} = \frac{1}{2}\left[\left(\epsilon^{(2)}|\widehat{A}|J^{-1}\right)_{j,k} + \left(\epsilon^{(2)}|\widehat{A}|J^{-1}\right)_{j+1,k}\right] \tag{4.81}$$

$$\left(\epsilon^{(4)}|\widehat{A}|J^{-1}\right)_{j+1/2,k} = \frac{1}{2}\left[\left(\epsilon^{(4)}|\widehat{A}|J^{-1}\right)_{j,k} + \left(\epsilon^{(4)}|\widehat{A}|J^{-1}\right)_{j+1,k}\right], \tag{4.82}$$

or a Roe average can be used for $\widehat{A}_{j+1/2,k}$.

The contribution of the second-difference term is controlled by a pressure sensor that detects shock waves [6, 12]. It is defined as follows:

$$\begin{aligned}
\epsilon^{(2)}_{j,k} &= \kappa_2 \max(\Upsilon_{j+1,k}, \Upsilon_{j,k}, \Upsilon_{j-1,k}) \\
\Upsilon_{j,k} &= \left|\frac{p_{j+1,k} - 2p_{j,k} + p_{j-1,k}}{p_{j+1,k} + 2p_{j,k} + p_{j-1,k}}\right| \\
\epsilon^{(4)}_{j,k} &= \max(0, \kappa_4 - \epsilon^{(2)}_{j,k}),
\end{aligned} \tag{4.83}$$

where typical values of the constants are $\kappa_2 = 1/2$ and $\kappa_4 = 1/50$. The switch is based on a normalized undivided second difference of pressure, which is much larger at shock waves than in smooth regions. The logic turns off the fourth-difference dissipation when the second-difference coefficient is large. The max function spreads out the contribution of the second-difference dissipation to ensure that it is not switched off in the interior of the shock.

Consistent with (4.76), the dissipative term can be written as

$$(D_\xi)_{j,k} = (d_\xi)_{j+1/2,k} - (d_\xi)_{j-1/2,k}, \tag{4.84}$$

where

$$(d_\xi)_{j+1/2,k} = \left(\epsilon^{(2)}|\widehat{A}|J^{-1}\right)_{j+1/2,k} \Delta_\xi Q_{j,k} - \left(\epsilon^{(4)}|\widehat{A}|J^{-1}\right)_{j+1/2,k} \Delta_\xi \nabla_\xi \Delta_\xi Q_{j,k}. \tag{4.85}$$

This ensures that the dissipation is conservative.

In order to reduce the cost of the dissipation model, one can replace the matrix $|\widehat{A}|$ with the spectral radius of $\widehat{A}$, which is its largest eigenvalue by absolute value. The spectral radius of $\widehat{A}$ is given by

$$\sigma = |U| + a\sqrt{\xi_x^2 + \xi_y^2}. \tag{4.86}$$

The spectral radius of $\widehat{B}$ is used for the η dissipation term. This approach, known as scalar artificial dissipation, leads to an inexpensive artificial dissipation scheme that is robust but can be excessively dissipative in certain contexts.

The astute reader may be wondering where the Δx terms in (4.66) and (4.67) have gone. These are implicit in $\widehat{A}$ and the spectral radius σ through the metric terms ξ_x and ξ_y, which scale with the inverse of the mesh spacing. This ensures that the two dissipation operators in (4.79) are first order and third order as desired.

Let us consider the fourth-difference dissipation term in more detail. Temporarily ignoring the coefficient term, we have

$$\begin{aligned}(D_\xi^{(4)})_{j,k} &= -\nabla_\xi \Delta_\xi \nabla_\xi \Delta_\xi Q_{j,k} \\ &= -Q_{j-2,k} + 4Q_{j-1,k} - 6Q_{j,k} + 4Q_{j+1,k} - Q_{j+2}.\end{aligned} \tag{4.87}$$

This operator involves values of Q from $j-2, k$ to $j+2, k$, i.e. a five-point stencil, in contrast to the finite-difference approximations to the inviscid and viscous flux derivatives, which involve data from $j-1$ to $j+1$ only, i.e. a three-point stencil. As we shall see in Sect. 4.5, this has significant implications for an implicit time-marching method. Here we are concerned with its implications near the boundaries of the grid. Boundary conditions are discussed later in this chapter. For now we will assume that the values of Q at the boundary are known, so the governing equations are not solved at the boundary. At the first interior node, the three-point operators for the inviscid and viscous fluxes as well as the second-difference dissipation can be applied without modification. However, the five-point operator cannot be applied, as either $Q_{j-2,k}$ or $Q_{j+2,k}$ is unavailable, depending on the boundary.

In developing a boundary scheme for the fourth-difference dissipation operator, one must ensure that the resulting scheme is conservative, dissipative, stable, and sufficiently accurate globally. First, we will consider conservation. The operator in (4.87) can be rewritten as

$$(D_\xi^{(4)})_{j,k} = (d_\xi^{(4)})_{j+1/2,k} - (d_\xi^{(4)})_{j-1/2,k}, \tag{4.88}$$

where

$$(d_\xi^{(4)})_{j+1/2,k} = Q_{j-1,k} - 3Q_{j,k} + 3Q_{j+1,k} - Q_{j+2,k}. \tag{4.89}$$

Without loss of generality, we will consider a boundary located at $j = 0$. The operator at $j = 1$ must be modified because the node $j - 2$ does not exist. Since the operator at $j = 2$ is not modified, conservation dictates that the term $(d_\xi^{(4)})_{j+1/2,k}$ at $j = 1$ cannot be modified. In any case, this term does not involve $Q_{j-2,k}$, so it need not be modified. There are several different ways to proceed; one is to define $(d_\xi^{(4)})_{j-1/2,k}$ at node $j = 1$ as

$$(d_\xi^{(4)})_{j-1/2,k} = -Q_{j-1,k} + 2Q_{j,k} - Q_{j+1,k}. \tag{4.90}$$

This leads to the following operator for the node at $j = 1$:

$$\begin{aligned}(D_\xi^{(4)})_{j,k} &= (Q_{j-1,k} - 3Q_{j,k} + 3Q_{j+1,k} - Q_{j+2,k}) \\ &\quad -(-Q_{j-1,k} + 2Q_{j,k} - Q_{j+1,k}) \\ &= 2Q_{j-1,k} - 5Q_{j,k} + 4Q_{j+1,k} - Q_{j+2,k}.\end{aligned} \tag{4.91}$$

Similar formulas are used at other boundaries. This approach has been shown to be dissipative and stable [14] and is therefore popular, although other options are also used. This boundary operator is first-order accurate locally and consistent with second-order global accuracy. If the interior scheme has an order of accuracy greater than two, then a higher order boundary operator should be used for the fourth-difference dissipation. Similarly, if better than third-order global accuracy is desired, then an artificial dissipation scheme of higher order is needed.

We conclude this section with a brief discussion of the application of this artificial dissipation scheme to the quasi-one-dimensional Euler equations, which are the subject of the exercises at the end of this chapter. The problems are to be solved on a uniform grid using the scalar artificial dissipation scheme. The spectral radius of the one-dimensional flux Jacobian matrix is

$$\sigma = |u| + a. \tag{4.92}$$

Since the mesh is uniform, no coordinate transformation is needed. The dissipation terms thus become

$$\begin{aligned}D_j &= \frac{1}{\Delta x}\nabla\left(\epsilon^{(2)}(|u| + a)\right)_{j+1/2}\Delta Q_j \\ &\quad -\frac{1}{\Delta x}\nabla\left(\epsilon^{(4)}(|u| + a)\right)_{j+1/2}\Delta\nabla\Delta Q_j,\end{aligned} \tag{4.93}$$

where ∇ and Δ denote undivided differences. Note in particular the $1/\Delta x$ scaling.

4.5 Implicit Time Marching and the Approximate Factorization Algorithm

After application of the above spatial discretization to (4.18), we obtain the following semi-discrete equation at each interior node in the mesh:

$$\partial_\tau \widehat{Q} = -\delta_\xi \widehat{E} + D_\xi - \delta_\eta \widehat{F} + D_\eta + Re^{-1}[\delta_\xi \widehat{E}_\mathrm{v} + \delta_\eta \widehat{F}_\mathrm{v}], \tag{4.94}$$

where δ represents the spatial difference operator, in this case second-order centered differences, and D_ξ and D_η the artificial dissipation terms, e.g. (4.79). Collecting these into a single equation, we obtain the following coupled system of nonlinear ODEs:

$$\frac{\mathrm{d}\widehat{\mathbf{Q}}}{\mathrm{d}t} = \mathbf{R}(\widehat{\mathbf{Q}}), \tag{4.95}$$

where $\widehat{\mathbf{Q}}$ is a column matrix containing $\widehat{Q}_{j,k}$ at each node of the mesh, $\mathbf{R}$ is a column matrix containing $R_{j,k}$ at each node, where

$$R(\widehat{Q}) = -\delta_\xi \widehat{E} + D_\xi - \delta_\eta \widehat{F} + D_\eta + Re^{-1}[\delta_\xi \widehat{E}_\mathrm{v} + \delta_\eta \widehat{F}_\mathrm{v}], \tag{4.96}$$

and we have replaced τ with t. In order to obtain a time-accurate solution for an unsteady flow problem, this system of ODEs must be solved using a time-marching method. Alternatively, if the flow under consideration is steady, one seeks the solution to the following coupled system of nonlinear algebraic equations:

$$\mathbf{R}(\widehat{\mathbf{Q}}) = \mathbf{0}. \tag{4.97}$$

In the steady case, $\mathbf{R}(\widehat{\mathbf{Q}})$ is referred to as the residual vector, or simply the residual. As a result of the nonlinear nature of the residual vector, this system cannot be solved directly: an iterative method is required.

For the numerical solution of a large system of nonlinear algebraic equations such as (4.97), it is natural to consider the Newton method, which produces the following linear system:

$$\mathbf{A}_n \Delta \widehat{\mathbf{Q}}_n = -\mathbf{R}(\widehat{\mathbf{Q}}_n), \tag{4.98}$$

where

$$\mathbf{A}_n = \frac{\partial \mathbf{R}}{\partial \widehat{\mathbf{Q}}} \tag{4.99}$$

is the Jacobian evaluated at state $\widehat{\mathbf{Q}}_n$, and $\Delta \widehat{\mathbf{Q}} = \widehat{\mathbf{Q}}_{n+1} - \widehat{\mathbf{Q}}_n$. This linear system must be solved iteratively until a converged solution is obtained that satisfies (4.97).

The degree to which a given iterate $\widehat{\mathbf{Q}}_n$ is a solution to (4.97) can be measured through the norm of $\mathbf{R}(\widehat{\mathbf{Q}})$. In finite precision arithmetic, it is typically not possible to reduce the norm of the residual below machine zero, so a solution for which this norm is on the order of machine zero can be considered fully converged. However, with single precision arithmetic, this level of convergence may not be sufficient.

Application of the Newton method to the large systems of nonlinear algebraic equations arising from the spatial discretization of the Euler or Navier-Stokes equations in multiple dimensions leads to two principal challenges. First, the Newton method converges only from an iterate that is within a finite region of convergence near the solution. Typically, the initial guess for $\widehat{\mathbf{Q}}$ lies outside this region, and some sort of globalization technique is needed to ensure that the Newton method will converge for an arbitrary initial iterate. A uniform flow is often used as the initial iterate. Second, the linear system of equations (4.98) that must be solved is in general large and sparse. Direct solution of such systems based on a lower-upper (LU) factorization can require a large amount of memory relative to the original sparse system and a number of floating point operations that scales poorly as the system size increases. Hence direct solution is only effective for linear systems below a certain size, although the system size for which direct solution of the system is a feasible approach increases with each new generation of computer hardware. The high cost of direct solution of this linear system for problems of practical interest motivates *inexact Newton methods* in which the linear system (4.98) is instead solved iteratively to some tolerance at each iteration. Sequences of tolerances can be found that maintain the quadratic convergence property of the Newton method within the radius of convergence, provided the residual function meets certain conditions.

A natural way to address the problem that the initial iterate is likely outside the region of convergence of the Newton method is to consider a time-dependent path to steady state. Under certain conditions, the solution of the steady problem (4.97) is also the steady solution of the ODE system (4.95), which can be found by applying a time-marching method to (4.95) and advancing in time until a steady state is reached. Time accuracy is not required; we simply wish to integrate in time from some arbitrary initial state to the steady solution in a manner that will require the smallest amount of computational work. The entire transient portion of the solution can be considered parasitic, and hence the problem is stiff. This suggests the use of an implicit time-marching method, and, given that we are not interested in time resolution of the transient, there is no reason to seek better than first-order accuracy. Therefore the implicit Euler method is the logical choice for steady problems. Its relationship with the Newton method is discussed in Sect. 2.6.3.

For unsteady flow problems where time-accurate solutions are required, one would like at least second-order accuracy. Hence, the trapezoidal and second-order backward methods (see Sect. 2.6), which are both unconditionally stable, are reasonable choices. The second-order backward method has a larger region of stability than the trapezoidal method, making it the more robust of the two. Moreover, the trapezoidal method provides little damping of modes with eigenvalues with large negative real parts, which is undesirable for stiff problems. Implicit Runge-Kutta methods, which we will not discuss here, are another option for time-accurate solution of stiff ODEs.

This brings us to the challenge of solving a large sparse linear system, which is present whether one is solving steady or unsteady problems. Historically, due to computer hardware limitations, direct solution techniques were not practical even for relatively small problems. Even today they are not an efficient option for large-scale three-dimensional problems. Inexact Newton methods have gained in popularity since the introduction of efficient iterative techniques for nonsymmetric sparse linear systems, such as the generailized minimal residual method (GMRES) [15]. However, these were not available until the mid-1980s, so the first implicit computations of three-dimensional flows were performed using the now classical approximate factorization algorithm, which is the subject of Sect. 4.5.4.

4.5.1 Implicit Time-Marching

Based on the above discussion, whether we are solving an unsteady problem or a steady one, we seek to solve the coupled system of ODEs given by (4.95) using an implicit time-marching method. We will consider the following two-parameter family of time-marching methods [3]:

$$\widehat{\mathbf{Q}}^{n+1} = \frac{\theta \Delta t}{1+\varphi}\frac{\mathrm{d}}{\mathrm{d}t}\widehat{\mathbf{Q}}^{n+1} + \frac{(1-\theta)\Delta t}{1+\varphi}\frac{\mathrm{d}}{\mathrm{d}t}\widehat{\mathbf{Q}}^{n} + \frac{1+2\varphi}{1+\varphi}\widehat{\mathbf{Q}}^{n} - \frac{\varphi}{1+\varphi}\widehat{\mathbf{Q}}^{n-1}$$
$$+O\left[(\theta - \frac{1}{2} - \varphi)\Delta t^2 + \Delta t^3\right], \tag{4.100}$$

where $\widehat{\mathbf{Q}}^n = \widehat{\mathbf{Q}}(n\Delta t)$. This family of methods is a subset of two-step linear multistep methods with the coefficient of

$$\frac{\mathrm{d}}{\mathrm{d}t}\widehat{\mathbf{Q}}^{n-1} \tag{4.101}$$

set to zero. One member of the family is third-order accurate, but that method is not of interest here, as it is not unconditionally stable. Our interest is in the first-order implicit Euler method obtained with $\theta = 1$ and $\varphi = 0$ for steady problems and the second-order backward method obtained with $\theta = 1$ and $\varphi = 1/2$ when time-accuracy is required.

For this exposition we will restrict ourselves to the implicit Euler method, but all of the subsequent development can easily be extended to any second-order scheme formed from (4.100). Applying the implicit Euler method to the thin-layer form of (4.95) results in the following expression at each node of the grid:

$$\widehat{Q}^{n+1} - \widehat{Q}^{n} = h\left(-\delta_\xi \widehat{E}^{n+1} + D_\xi^{n+1} - \delta_\eta \widehat{F}^{n+1} + D_\eta^{n+1} + Re^{-1}\delta_\eta \widehat{S}^{n+1}\right), \tag{4.102}$$

with $h = \Delta t$.

4.5.2 Local Time Linearization

We wish to solve (4.102) for $\widehat{Q}^{n+1}$ given $\widehat{Q}^n$. The flux vectors $\widehat{E}$, $\widehat{F}$, and $\widehat{S}$, and the artificial dissipation terms D_ξ, and D_η, are nonlinear functions of $\widehat{Q}$, and therefore the right-hand side of (4.102) is nonlinear in $\widehat{Q}^{n+1}$. Hence we proceed by locally linearizing with respect to t.

The flux vectors are linearized in time about $\widehat{Q}^n$ by Taylor series such that

$$\begin{aligned}
\widehat{E}^{n+1} &= \widehat{E}^n + \widehat{A}^n \Delta\widehat{Q}^n + O(h^2)\\
\widehat{F}^{n+1} &= \widehat{F}^n + \widehat{B}^n \Delta\widehat{Q}^n + O(h^2)\\
Re^{-1}\widehat{S}^{n+1} &= Re^{-1}\left[\widehat{S}^n + \widehat{M}^n \Delta\widehat{Q}^n\right] + O(h^2),
\end{aligned} \tag{4.103}$$

where $\widehat{A} = \partial\widehat{E}/\partial\widehat{Q}$, $\widehat{B} = \partial\widehat{F}/\partial\widehat{Q}$ and $\widehat{M} = \partial\widehat{S}/\partial\widehat{Q}$ are the flux Jacobians, and $\Delta\widehat{Q}^n$ is $O(h)$. As discussed in Sect. 2.6.3, such a local time linearization will not degrade the order of accuracy of time-marching methods of up to second order.

The inviscid flux Jacobian matrices $\widehat{A}$ and $\widehat{B}$ are given by

$$\begin{bmatrix}
\kappa_t & \kappa_x & \kappa_y & 0\\
-u\theta + \kappa_x\phi^2 & \kappa_t + \theta - (\gamma-2)\kappa_x u & \kappa_y u - (\gamma-1)\kappa_x v & (\gamma-1)\kappa_x\\
-v\theta + \kappa_y\phi^2 & \kappa_x v - (\gamma-1)\kappa_y u & \kappa_t + \theta - (\gamma-2)\kappa_y v & (\gamma-1)\kappa_y\\
\theta[\phi^2 - a_1] & \kappa_x a_1 - (\gamma-1)u\theta & \kappa_y a_1 - (\gamma-1)v\theta & \gamma\theta + \kappa_t
\end{bmatrix}, \tag{4.104}$$

with $a_1 = \gamma(e/\rho) - \phi^2$, $\theta = \kappa_x u + \kappa_y v$, $\phi^2 = \frac{1}{2}(\gamma-1)(u^2+v^2)$, and $\kappa = \xi$ or η for $\widehat{A}$ or $\widehat{B}$, respectively. As an example, we will derive the first element in the second row of $\widehat{A}$, i.e.

$$\widehat{a}_{21} = \frac{\partial\widehat{e}_2}{\partial\widehat{q}_1}, \tag{4.105}$$

where

$$\widehat{Q} = \begin{bmatrix}\widehat{q}_1\\ \widehat{q}_2\\ \widehat{q}_3\\ \widehat{q}_4\end{bmatrix} = J^{-1}\begin{bmatrix}\rho\\ \rho u\\ \rho v\\ e\end{bmatrix}, \quad \widehat{E} = \begin{bmatrix}\widehat{e}_1\\ \widehat{e}_2\\ \widehat{e}_3\\ \widehat{e}_4\end{bmatrix} = J^{-1}\begin{bmatrix}\rho U\\ \rho u U + \xi_x p\\ \rho v U + \xi_y p\\ U(e+p) - \xi_t p\end{bmatrix}. \tag{4.106}$$

In order to find $\widehat{a}_{21}$, the first step is to write $\widehat{e}_2$ in terms of the elements of $\widehat{Q}$. One obtains

$$
\begin{aligned}
\widehat{e}_2 &= J^{-1}\rho u U + J^{-1}\xi_x p \\
&= J^{-1}\rho u \xi_t + J^{-1}\rho u^2 \xi_x + J^{-1}\rho u v \xi_y \\
&\quad + J^{-1}\xi_x(\gamma-1)e - J^{-1}\xi_x(\gamma-1)\frac{1}{2}\rho u^2 - J^{-1}\xi_x(\gamma-1)\frac{1}{2}\rho v^2 \\
&= \xi_t \widehat{q}_2 + \xi_x \frac{\widehat{q}_2^2}{\widehat{q}_1} + \xi_y \frac{\widehat{q}_2\widehat{q}_3}{\widehat{q}_1} + \xi_x(\gamma-1)\widehat{q}_4 - \frac{\xi_x(\gamma-1)}{2}\frac{\widehat{q}_2^2}{\widehat{q}_1} - \frac{\xi_x(\gamma-1)}{2}\frac{\widehat{q}_3^2}{\widehat{q}_1}.
\end{aligned}
\tag{4.107}
$$

From this we find

$$
\begin{aligned}
\widehat{a}_{21} = \frac{\partial \widehat{e}_2}{\partial \widehat{q}_1} &= -\xi_x \frac{\widehat{q}_2^2}{\widehat{q}_1^2} - \xi_y \frac{\widehat{q}_2\widehat{q}_3}{\widehat{q}_1^2} + \frac{\xi_x(\gamma-1)}{2}\frac{\widehat{q}_2^2}{\widehat{q}_1^2} + \frac{\xi_x(\gamma-1)}{2}\frac{\widehat{q}_3^2}{\widehat{q}_1^2} \\
&= -\xi_x u^2 - \xi_y u v + \frac{\xi_x(\gamma-1)}{2}u^2 + \frac{\xi_x(\gamma-1)}{2}v^2 \\
&= -u(\xi_x u + \xi_y v) + \frac{\xi_x(\gamma-1)}{2}(u^2+v^2),
\end{aligned}
\tag{4.108}
$$

consistent with (4.104). The other terms in $\widehat{A}$ and $\widehat{B}$ are found in a similar manner.

The thin-layer viscous flux Jacobian is

$$
\widehat{M} = J^{-1}\begin{bmatrix} 0 & 0 & 0 & 0 \\ m_{21} & \alpha_1\partial_\eta(\rho^{-1}) & \alpha_2\partial_\eta(\rho^{-1}) & 0 \\ m_{31} & \alpha_2\partial_\eta(\rho^{-1}) & \alpha_3\partial_\eta(\rho^{-1}) & 0 \\ m_{41} & m_{42} & m_{43} & m_{44} \end{bmatrix} J,
\tag{4.109}
$$

where

$$
\begin{aligned}
m_{21} &= -\alpha_1\partial_\eta(u/\rho) - \alpha_2\partial_\eta(v/\rho) \\
m_{31} &= -\alpha_2\partial_\eta(u/\rho) - \alpha_3\partial_\eta(v/\rho) \\
m_{41} &= \alpha_4\partial_\eta\left[-(e/\rho^2) + (u^2+v^2)/\rho\right] \\
&\quad -\alpha_1\partial_\eta(u^2/\rho) - 2\alpha_2\partial_\eta(uv/\rho) \\
&\quad -\alpha_3\partial_\eta(v^2/\rho) \\
m_{42} &= -\alpha_4\partial_\eta(u/\rho) - m_{21} \\
m_{43} &= -\alpha_4\partial_\eta(v/\rho) - m_{31} \\
m_{44} &= \alpha_4\partial_\eta(\rho^{-1}) \\
\alpha_1 &= \mu[(4/3){\eta_x}^2 + {\eta_y}^2], \quad \alpha_2 = (\mu/3)\eta_x\eta_y \\
\alpha_3 &= \mu[{\eta_x}^2 + (4/3){\eta_y}^2], \quad \alpha_4 = \gamma\mu Pr^{-1}({\eta_x}^2 + {\eta_y}^2).
\end{aligned}
$$

Its derivation is made more complicated by virtue of the fact that $\widehat{S}$ includes within it derivatives of $\widehat{Q}$. Therefore the term $\widehat{M}^n \Delta \widehat{Q}^n$ in (4.103) also must contain derivatives of $\Delta \widehat{Q}^n$, so this term is not a simple matrix-vector product as is the case for the terms $\widehat{A}^n \Delta \widehat{Q}^n$ and $\widehat{B}^n \Delta \widehat{Q}^n$.

To clarify this, let us derive the second element in the second row of $\widehat{M}$. We begin by writing the second element of $\widehat{S}$ in terms of $\widehat{Q}$ as follows:

$$\begin{aligned}\widehat{s}_2 &= \frac{\alpha_1}{J} u_\eta + \frac{\alpha_2}{J} v_\eta \\ &= \frac{\alpha_1}{J} \frac{\partial}{\partial \eta}\left(\frac{\widehat{q}_2}{\widehat{q}_1}\right) + \frac{\alpha_2}{J} \frac{\partial}{\partial \eta}\left(\frac{\widehat{q}_3}{\widehat{q}_1}\right),\end{aligned} \tag{4.110}$$

where α_1 and α_2 are defined below (4.109). For this derivation we retain the analytical derivative from the original PDE rather than the finite-difference approximation, which can be applied later. It is clear that the second term on the right-hand side in (4.110), which does not involve $\widehat{q}_2$, will not enter into the term $\widehat{m}_{22}$ in $\widehat{M}$. Hence we define an operator $f(\widehat{q}_2)$ as follows:

$$f(\widehat{q}_2) = \frac{\alpha_1}{J} \frac{\partial}{\partial \eta}\left(\frac{\widehat{q}_2}{\widehat{q}_1}\right), \tag{4.111}$$

which is the first term in (4.110). We can then use a Fréchet derivative to find

$$\begin{aligned}\frac{\partial f}{\partial \widehat{q}_2} \Delta \widehat{q}_2 &= \lim_{\epsilon \to 0} \frac{f(\widehat{q}_2 + \epsilon \Delta \widehat{q}_2) - f(\widehat{q}_2)}{\epsilon} \\ &= \lim_{\epsilon \to 0} \left[\frac{\alpha_1}{J} \frac{\partial}{\partial \eta}\left(\frac{\widehat{q}_2 + \epsilon \Delta \widehat{q}_2}{\widehat{q}_1}\right) - \frac{\alpha_1}{J} \frac{\partial}{\partial \eta}\left(\frac{\widehat{q}_2}{\widehat{q}_1}\right)\right] / \epsilon \\ &= \lim_{\epsilon \to 0} \left[\frac{\alpha_1}{J} \frac{\partial}{\partial \eta}\left(\frac{\epsilon \Delta \widehat{q}_2}{\widehat{q}_1}\right)\right] / \epsilon \\ &= \frac{\alpha_1}{J} \frac{\partial}{\partial \eta}\left(\frac{\Delta \widehat{q}_2}{\widehat{q}_1}\right).\end{aligned} \tag{4.112}$$

Thus we see that the product $\widehat{m}_{22} \Delta \widehat{q}_2$ is

$$\widehat{m}_{22} \Delta \widehat{q}_2 = J^{-1} \alpha_1 \frac{\partial}{\partial \eta}\left(\frac{J}{\rho} \Delta \widehat{q}_2\right). \tag{4.113}$$

This is identical to (4.109) and clarifies the precise meaning of that equation. The ∂_η derivatives in $\widehat{M}$ operate on the product of the term shown in $\widehat{M}$, e.g. ρ^{-1} in $\widehat{m}_{22}$, the J term shown to the right of the matrix in (4.109), and the appropriate component of $\Delta \widehat{Q}$.

The nonlinear artificial dissipation terms D_ξ and D_η appearing in (4.102) must also be locally linearized. As a result of the complexity of (4.79), for example, an inexact linearization of these terms is often used, especially in the context of the approximate factorization algorithm. This is achieved by treating the coefficient terms in the artificial dissipation, such as $\epsilon^{(4)}|\widehat{A}|$ in (4.79), as frozen at time level n, making the linearization straightforward. This approximation is not made on the right-hand side.

Substituting the local time linearizations of the nonlinear flux vectors in (4.103) into (4.102) and grouping the $\Delta\widehat{Q}^n$ terms on the left-hand side produces the *delta form* of the algorithm:

$$\begin{aligned}\left[I + h\delta_\xi\widehat{A}^n - hL_\xi + h\delta_\eta\widehat{B}^n - hL_\eta - Re^{-1}h\,\delta_\eta\widehat{M}\right]\Delta\widehat{Q}^n \\ = -h\left(\delta_\xi\widehat{E}^n - D_\xi^n + \delta_\eta\widehat{F}^n - D_\eta^n - Re^{-1}\delta_\eta\widehat{S}^n\right),\end{aligned} \tag{4.114}$$

where L_ξ and L_η result from the linearization of the artificial dissipation terms. The right-hand side is simply h times the right-hand side of the thin-layer form of (4.94). This results in an important property of the delta form. If a fully converged steady solution of (4.114) is obtained, then it will be the correct steady solution of (4.94), *independent of the left-hand side of* (4.114). This means that approximations made to the left-hand side in order to reduce the computational work needed to converge to steady state, i.e. to drive the norm of $\mathbf{R}(\widehat{\mathbf{Q}})$ to machine zero, will have no effect on the converged solution.

The finite-difference operators on the left-hand side of (4.114) operate on the product of the terms immediately to their right within the square brackets and the $\Delta\widehat{Q}^n$ outside the square brackets. For example, the δ_ξ term results in

$$\frac{1}{2}h(\widehat{A}^n_{j+1,k}\Delta\widehat{Q}^n_{j+1,k} - \widehat{A}^n_{j-1,k}\Delta\widehat{Q}^n_{j-1,k}). \tag{4.115}$$

The viscous contribution on the left-hand side includes both the δ_η term shown in (4.114) and the finite-difference approximations of the partial derivatives with respect to η within the viscous flux Jacobian $\widehat{M}$. These must be consistent with the compact three-point operator used on the right-hand side given in (4.48). The $\Delta\widehat{Q}^n$ terms are of course unknown, and (4.114) represents a linear system of equations to be solved at each iteration of the implicit Euler method. Excluding the I term, the terms within the square brackets on the left-hand side of (4.114) are a linearization of the negative discrete residual operator, i.e. the negative of the right-hand side. Consequently, if the I term is omitted, we obtain the Newton method, consistent with the fact that the Newton method is obtained from the time linearized implicit Euler method in the limit as h goes to infinity (see Sect. 2.6.3).

4.5.3 Matrix Form of the Unfactored Algorithm

We refer to (4.114) as the unfactored algorithm. It produces a large banded system of algebraic equations. We now examine the associated matrix. Let the number of grid nodes in the ξ direction be J and in the η direction K. Temporarily ignoring the viscous and artificial dissipation terms, the banded matrix is a $(J \cdot K \cdot 4) \times (J \cdot K \cdot 4)$ square matrix of the form

$$
\left[I + h\delta_\xi \widehat{A}^n + h\delta_\eta \widehat{B}^n \right] \Rightarrow
$$

$$
\left[\begin{array}{cccccccccccc}
I & h\widehat{A}/2 & & & h\widehat{B}/2 & & & & & & & \\
-h\widehat{A}/2 & I & h\widehat{A}/2 & & & h\widehat{B}/2 & & & & & & \\
& -h\widehat{A}/2 & I & h\widehat{A}/2 & & & h\widehat{B}/2 & & & & & \\
& & -h\widehat{A}/2 & I & h\widehat{A}/2 & & & h\widehat{B}/2 & & & & \\
-h\widehat{B}/2 & & & -h\widehat{A}/2 & I & h\widehat{A}/2 & & & h\widehat{B}/2 & & & \\
& \ddots & & & \ddots & \ddots & \ddots & & & \ddots & & \\
& & -h\widehat{B}/2 & & & -h\widehat{A}/2 & I & h\widehat{A}/2 & & & h\widehat{B}/2 & \\
& & & -h\widehat{B}/2 & & & -h\widehat{A}/2 & I & h\widehat{A}/2 & & & h\widehat{B}/2 \\
& & & & -h\widehat{B}/2 & & & -h\widehat{A}/2 & I & h\widehat{A}/2 & & \\
& & & & & \ddots & & & \ddots & \ddots & \ddots & \\
& & & & & & -h\widehat{B}/2 & & & -h\widehat{A}/2 & I & h\widehat{A}/2 \\
& & & & & & & -h\widehat{B}/2 & & & -h\widehat{A}/2 & I
\end{array} \right], \qquad (4.116)
$$

where the variables have been ordered with j running first and then k. Each entry is a 4×4 block. If we order the variables with k running first and then j, the roles of $\widehat{A}$ and $\widehat{B}$ are reversed in the above matrix, i.e. the $h\widehat{B}$ terms produce a tridiagonal form, while the $h\widehat{A}$ terms produce a much larger bandwidth. The thin-layer viscous terms involve a three-point operator in the η direction, so they add to the diagonal block and contribute to the $h\widehat{B}$ blocks shown in (4.116), but they do not alter the overall structure of the matrix. Finally, the artificial dissipation terms involve a five-point operator in each direction and thus further increase the matrix bandwidth. If the scalar artificial dissipation model is used, the corresponding entries are in the form σI, where σ is a scalar, and I is the 4×4 identity matrix.

Although this matrix is sparse, it would be very expensive computationally to solve the algebraic system directly through an LU factorization. For example, for an accurate computation of a three-dimensional transonic flow past a wing, one can easily require over ten million mesh nodes. The resulting linear system is a 50 million $\times$ 50 million matrix problem to be solved, and although one could take advantage of its banded sparse structure, it would still be very costly in terms of both computational work and memory. This motivates iterative and approximate solution strategies for sparse linear systems, such as the approximate factorization algorithm described next.

4.5.4 Approximate Factorization

One way to reduce the computational cost of the solution process is to introduce an approximate factorization of the two-dimensional operator into two one-dimensional operators. Ignoring the artificial dissipation for now, the left-hand side of (4.114) can be written as

$$\left[I + h\delta_\xi \widehat{A}^n + h\delta_\eta \widehat{B}^n - hRe^{-1}\delta_\eta \widehat{M}^n\right] \Delta\widehat{Q}^n$$
$$= \left[I + h\delta_\xi \widehat{A}^n\right] \left[I + h\delta_\eta \widehat{B}^n - hRe^{-1}\delta_\eta \widehat{M}^n\right] \Delta\widehat{Q}^n$$
$$-h^2\delta_\xi \widehat{A}^n \delta_\eta \widehat{B}^n \Delta\widehat{Q}^n + h^2 Re^{-1}\delta_\xi \widehat{A}^n \delta_\eta \widehat{M}^n \Delta\widehat{Q}^n. \tag{4.117}$$

Noting that $\Delta\widehat{Q}^n$ is $O(h)$, the difference between the factored form and the unfactored form is $O(h^3)$. Therefore, this difference can be neglected without reducing the time accuracy below second order.

The resulting factored form of the algorithm is

$$\left[I + h\delta_\xi \widehat{A}^n\right]\left[I + h\delta_\eta \widehat{B}^n - hRe^{-1}\delta_\eta \widehat{M}^n\right] \Delta\widehat{Q}^n \tag{4.118}$$
$$= -h\left[\delta_\xi \widehat{E}^n + \delta_\eta \widehat{F}^n - Re^{-1}\delta_\eta \widehat{S}^n\right].$$

We now have two matrices each of which is block tridiagonal if the appropriate ordering of the variables is used. The structure of the block tridiagonal matrices is

$$\left[I + h\delta_\xi \widehat{A}^n\right] \Rightarrow \begin{bmatrix} I & h\widehat{A}/2 & & & & & & \\ -h\widehat{A}/2 & I & h\widehat{A}/2 & & & & & \\ & -h\widehat{A}/2 & I & h\widehat{A}/2 & & & & \\ & & \ddots & \ddots & \ddots & & & \\ & & & -h\widehat{A}/2 & I & h\widehat{A}/2 & & \\ & & & & -h\widehat{A}/2 & I & h\widehat{A}/2 & \\ & & & & & -h\widehat{A}/2 & I & h\widehat{A}/2 \\ & & & & & & -h\widehat{A}/2 & I \end{bmatrix}.$$

The thin-layer viscous term $\widehat{M}$ is kept with the η factor. Since it is also based upon a three-point stencil, it will not affect the tridiagonal structure.

The mechanics of the approximate factorization algorithm are as follows. First solve the system

$$\left[I + h\delta_\xi \widehat{A}^n\right] \Delta\tilde{Q} = -h\left[\delta_\xi \widehat{E}^n + \delta_\eta \widehat{F}^n - Re^{-1}\delta_\eta \widehat{S}^n\right] \tag{4.119}$$

for $\Delta\tilde{Q}$, where $\Delta\tilde{Q}$ is an intermediate variable. This requires K solutions of a $(J\cdot 4)\times(J\cdot 4)$ system. With the variables ordered with j running first, followed by k, this is a block tridiagonal system, which can be efficiently solved by a block lower-upper (LU) decomposition. This step is equivalent to solving K one-dimensional problems, one for each ξ line in the mesh.

The next step is to permute, or reorder, $\Delta\tilde{Q}$ such that k is running first, followed by j. This reordering is only conceptual. In practice, this is handled through programming using array indices. Then solve

$$\left[I + h\delta_\eta \widehat{B}^n - hRe^{-1}\delta_\eta \widehat{M}^n\right] \Delta\widehat{Q}^n = \Delta\tilde{Q} \tag{4.120}$$

for $\Delta\widehat{Q}^n$. This requires J solutions of a $(K \cdot 4) \times (K \cdot 4)$ system. With the variables ordered with k running first, followed by j, this is also a block tridiagonal system. This step is equivalent to solving J one-dimensional problems, one for each η line in the mesh. The resulting vector $\Delta\widehat{Q}^n$ must be reordered back to the original database with j running first, again only conceptually, and added to $\widehat{Q}^n$ to form $\widehat{Q}^{n+1}$.

Since efficient specialized algorithms can be used to solve block tridiagonal systems, the factored form substantially reduces the computational work required for one implicit time step. Moreover, as a result of the use of the delta form, we are assured that the steady-state solution is unaffected by the factorization of the left-hand side operator. What remains to be seen is the effect of the factorization on the number of iterations needed to converge to the steady state. This we examine next.

For this purpose we will consider the following simple scalar model ODE:

$$\frac{du}{dt} = [\lambda_x + \lambda_y]u + a, \tag{4.121}$$

where λ_x, λ_y, and a are complex constants, which has the exact solution

$$u(t) = ce^{(\lambda_x+\lambda_y)t} - \frac{a}{\lambda_x + \lambda_y}. \tag{4.122}$$

We will assume that both λ_x and λ_y have negative real parts, so the ODE is inherently stable and has a steady solution given by

$$\lim_{t\to\infty} u(t) = -\frac{a}{\lambda_x + \lambda_y}. \tag{4.123}$$

Following the approach of Sect. 2.6.2, application of the unfactored form of the implicit Euler method leads to an OΔE that has the following solution:

$$u_n = c\sigma^n - \frac{a}{\lambda_x + \lambda_y}, \tag{4.124}$$

where

$$\sigma = \frac{1}{1 - h\,\lambda_x - h\,\lambda_y}.$$

This method is unconditionally stable and converges rapidly to the steady-state solution for large h because the magnitude of the amplification factor $|\sigma| \to 0$ as $h \to \infty$. As discussed, however, when applied to practical problems, the cost of this method can be prohibitive.

In contrast, the approximate factorization presented in this chapter produces the following OΔE when applied to (4.121):

$$(1 - h\,\lambda_x)\big(1 - h\,\lambda_y\big)(u_{n+1} - u_n) = h\big(\lambda_x u_n + \lambda_y u_n + a\big),$$

which reduces to

$$(1 - h\,\lambda_x)\big(1 - h\,\lambda_y\big)u_{n+1} = \Big(1 + h^2\lambda_x\lambda_y\Big)u_n + ha.$$

The solution of this OΔE is given by (4.124) with

$$\sigma = \frac{1 + h^2\lambda_x\lambda_y}{(1 - h\,\lambda_x)\big(1 - h\,\lambda_y\big)}. \tag{4.125}$$

Although this method remains unconditionally stable and produces the exact steady-state solution independent of h, it converges very slowly to the steady-state solution for large values of h, since the magnitude of the amplification factor $|\sigma| \to 1$ as $h \to \infty$. The factoring error has introduced an h^2 term in the numerator of the amplification factor that destroys the good convergence characteristics at large time steps. In comparison with the unfactored method, the factored form will take more iterations to converge, but each iteration will involve much less computational work.

Let us examine this in more detail. The amplification factor approaches unity as h goes to zero, and, for the factored form, its magnitude also tends to unity as h goes to infinity. The magnitude of the amplification factor thus has a minimum for some value of h, and this is the optimum choice of h for rapid convergence to steady-state. When solving a system of ODEs, there are many eigenvalues, and one cannot choose the optimum value of h for each one. Instead, one seeks an h that balances the magnitude of the amplification factor associated with the smallest eigenvalues with that associated with the largest eigenvalues. Choosing a smaller h will increase the amplification factor for the smallest eigenvalue, while a larger h will increase the amplification factor for the largest eigenvalue. Hence this choice of h is optimal in the sense that it minimizes the maximum amplification factor.

One can contrast this with the time step choice for an explicit time-marching scheme applied to a steady problem. Such schemes are conditionally stable, so there is firm upper bound on the time step. Optimal convergence to steady state is usually achieved with a time step just slightly below this stability limit. In other words, h must be chosen such that the largest eigenvalues lie in the stable region of the explicit method, which is generally a smaller time step than would be optimal for the factored implicit method. Therefore, the amplification factor for the smallest eigenvalues will be larger than for the factored method, and a larger number of iterations will be needed to reach a steady state. This must of course be weighed against the reduced cost per time step of the explicit method. As the spread in the eigenvalues increases, i.e. the problem becomes stiffer, the advantage tilts toward the implicit method. For example, implicit methods are typically preferred for problems involving chemical reactions or grid cells with very high aspect ratios as needed for the computation of turbulent flows at high Reynolds numbers.

Now we return to the contribution of the linearization of the artificial dissipation terms to the left-hand side of (4.114). The first operator, L_ξ, operates solely in the ξ direction, while the second, L_η, operates solely in the η direction. Hence these operators are amenable to approximate factorization with hL_ξ added to the $\left[I + h\delta_\xi \widehat{A}^n\right]$ factor and hL_η to the $\left[I + h\delta_\eta \widehat{B}^n - hRe^{-1}\delta_\eta \widehat{M}^n\right]$ factor. Since the artificial dissipation operators involve a five-point stencil, the matrices become block pentadiagonal rather than block tridiagonal.

4.5.5 Diagonal Form of the Implicit Algorithm

The approximate factorization algorithm based on solving block pentadiagonal factors is a viable and efficient algorithm. Nevertheless, the majority of the computational work resides in solving the block pentadiagonal systems, so it is worthwhile to examine strategies to reduce this. One way to reduce the computational work is to introduce a diagonalization of the blocks in the implicit operators, as developed by Pulliam and Chaussee [5]. The eigensystems of the flux Jacobians $\widehat{A}$ and $\widehat{B}$ are used in this construction. For now let us restrict ourselves to the Euler equations; application to the Navier-Stokes equations is discussed later.

The flux Jacobians $\widehat{A}$ and $\widehat{B}$ each have real eigenvalues and a complete set of eigenvectors. Therefore, the Jacobian matrices can be diagonalized as follows (see Warming et al. [16]):

$$\Lambda_\xi = T_\xi^{-1}\widehat{A}T_\xi \quad \text{and} \quad \Lambda_\eta = T_\eta^{-1}\widehat{B}T_\eta, \tag{4.126}$$

where Λ_ξ and Λ_η are diagonal matrices containing the eigenvalues of $\widehat{A}$ and $\widehat{B}$, T_ξ is a matrix whose columns are the eigenvectors of $\widehat{A}$, and T_η is the corresponding eigenvector matrix for $\widehat{B}$. These matrices are written out in the Appendix. We take the factored algorithm in delta form (4.118), neglect the viscous terms, and replace $\widehat{A}$ and $\widehat{B}$ with their respective eigensystem decompositions to obtain:

$$\begin{aligned}&\left[T_\xi\, T_\xi^{-1} + h\ \delta_\xi \left(T_\xi\, \Lambda_\xi\, T_\xi^{-1}\right)\right] \left[T_\eta\, T_\eta^{-1} + h\ \delta_\eta \left(T_\eta\, \Lambda_\eta\, T_\eta^{-1}\right)\right] \Delta\widehat{Q}^n \\ &\quad = -h\left[\delta_\xi \widehat{E}^n + \delta_\eta \widehat{F}^n\right] = \widehat{R}^n.\end{aligned} \tag{4.127}$$

Note that the identity matrix I has been replaced by $T_\xi T_\xi^{-1}$ and $T_\eta T_\eta^{-1}$ in each factor, respectively.

At this point, no approximations have been made, and with the exception of the viscous terms, (4.118) and (4.127) are equivalent. A modified form of (4.127) can be obtained by factoring the T_ξ and T_η eigenvector matrices outside the spatial derivative terms δ_ξ and δ_η. The eigenvector matrices are functions of ξ and η, and therefore this modification introduces an approximation on the left-hand side. The resulting equations are

$$T_\xi\left[I + h\,\delta_\xi\,\Lambda_\xi\right]\widehat{N}\left[I + h\ \delta_\eta\,\Lambda_\eta\right]T_\eta^{-1}\Delta\widehat{Q}^n = \widehat{R}^n, \tag{4.128}$$

where $\widehat{N} = T_\xi^{-1}T_\eta$ (see Appendix).

The approximation made to the left-hand side of (4.127) reduces the time accuracy to at best first order, and, moreover, gives time-accurate computations a nonconservative feature that leads to errors in shock speeds and jump conditions. However, the right-hand side is unmodified, so if the algorithm converges, it will converge to the correct steady-state solution. The advantage of the diagonal form is that the equations are decoupled as a result. Rather than a block tridiagonal system, we now have four scalar tridiagonal systems plus some additional 4×4 matrix-vector multiplies, leading to a substantial reduction in computational work. The computational work can be further decreased by exploiting the fact that the first two eigenvalues of the system are identical (see Appendix). This allows us to combine the coefficient calculations and part of the inversion work for the first two scalar operators.

The diagonal form reduces the computational work per time step and produces the correct steady solution. The next step is to examine its effect on the number of time steps needed to converge to steady state. Normally one would turn to linear stability analysis to assess the stability limits and convergence rate of an algorithm. However, linear analysis is of no use in analyzing the diagonal algorithm because the assumption of linear analysis is that the Jacobians are constant. With this assumption, the diagonalization introduces no approximation at all, so linear stability analysis predicts the diagonal algorithm to have the same unconditional stability as the original block algorithm. Therefore one must resort to computational experiments in order to investigate the impact of the diagonal form on the convergence properties of the diagonal algorithm. Pulliam and Chaussee [5] have shown that the convergence and stability limits of the diagonal algorithm are similar to those of the block form of the algorithm. The reader will have the opportunity to perform similar experiments as part of the exercises at the end of this chapter.

The steps involved in applying the diagonal form of the approximate factorization algorithm are as follows:

1. Beginning with (4.128), premultiply $\widehat{R}^n$ by T_ξ^{-1} to obtain the system

$$\left[I + h\,\delta_\xi\,\Lambda_\xi\right]\widehat{N}\left[I + h\,\delta_\eta\,\Lambda_\eta\right]T_\eta^{-1}\Delta\widehat{Q}^n = T_\xi^{-1}\widehat{R}^n. \tag{4.129}$$

2. With the variables ordered with j running first, solve the scalar trididagonal system

$$\left[I + h\,\delta_\xi\,\Lambda_\xi\right]X_1 = T_\xi^{-1}\widehat{R}^n \tag{4.130}$$

for the temporary variable X_1. This produces the following

$$\widehat{N}\left[I + h\,\delta_\eta\,\Lambda_\eta\right]T_\eta^{-1}\Delta\widehat{Q}^n = X_1. \tag{4.131}$$

3. Premultiply by $\widehat{N}^{-1}$ to obtain

$$\left[I + h\,\delta_\eta\,\Lambda_\eta\right]T_\eta^{-1}\Delta\widehat{Q}^n = \widehat{N}^{-1}X_1. \tag{4.132}$$

4. With the variables ordered with k running first, solve the scalar tridiagonal system

$$\left[I + h\,\delta_\eta\,\Lambda_\eta\right] X_2 = \widehat{N}^{-1} X_1 \tag{4.133}$$

for X_2, giving

$$T_\eta^{-1}\Delta\widehat{Q}^n = X_2. \tag{4.134}$$

5. Premultiply X_2 by T_η to find $\Delta\widehat{Q}^n$.

The diagonal algorithm as presented above is only strictly valid for the Euler equations. This is because we have neglected the implicit linearization of the viscous flux $\widehat{S}^n$ in the implicit operator for the η direction. The viscous flux Jacobian $\widehat{M}^n$ is not simultaneously diagonalizable with the inviscid flux Jacobian $\widehat{B}^n$ and therefore to include it in the diagonal form is not straightforward. For viscous flows one can consider four options. One possibility is to use the diagonal form in the ξ direction only and the block algorithm in the η direction. This increases the computational work substantially. Another option is to introduce a third factor to the implicit side of Eq. 4.118 as follows:

$$\left[I - hRe^{-1}\delta_\eta\widehat{M}^n\right]. \tag{4.135}$$

This again increases the computational work since we now have an added block tridiagonal inversion. One could diagonalize this term, but it would nevertheless increase the cost substantially. The third option is to throw caution to the wind and actually neglect the viscous Jacobian, thereby gaining the increased efficiency of the diagonal algorithm. This can have an adverse effect on stability and convergence. The fourth option is to include a diagonal term on the implicit side that is a rough approximation to the viscous Jacobian spectral radius. Estimates that have been used successfully are

$$\begin{aligned}
\lambda_v(\xi) &= \gamma Pr^{-1}\mu Re^{-1}\left(\xi_x^2 + \xi_y^2\right)\rho^{-1} \\
\lambda_v(\eta) &= \gamma Pr^{-1}\mu Re^{-1}\left(\eta_x^2 + \eta_y^2\right)\rho^{-1},
\end{aligned} \tag{4.136}$$

which are added to the appropriate operators in Eq. 4.128 with a differencing stencil taken from Eq. 4.48. With these terms added, the diagonal algorithm is given as

$$T_\xi\left[I + h\,\delta_\xi\,\Lambda_\xi - h\,I\,\delta_{\xi\xi}\lambda_v(\xi)\right]\widehat{N}\left[I + h\,\delta_\eta\,\Lambda_\eta - h\,I\,\delta_{\eta\eta}\lambda_v(\eta)\right]T_\eta^{-1}\Delta\widehat{Q}^n = \widehat{R}^n. \tag{4.137}$$

The ξ term is not added if the thin layer approximation is used. Although this approach is not rigorous, given that the eigenvectors of the viscous Jacobians are distinct from those of the inviscid Jacobians, it has proven to be effective in terms of both efficiency and reliability. It is thus the recommended approach for application of the diagonal form to viscous flows.

Next we consider the contribution of the linearization of the artificial dissipation terms, L_ξ and L_η in (4.114), in the context of the diagonal algorithm. Recall that the operator associated with the fourth-difference dissipation leads to a pentadiagonal matrix rather than a tridiagonal matrix, so the full block algorithm requires the solution of block pentadiagonal systems. If scalar dissipation is used, the contributions to the left-hand side are in the form σI, where σ is a scalar, so this is directly compatible with the diagonal form. With matrix dissipation, the diagonalization is also straightforward, since $\widehat{A}$ and $|\widehat{A}|$ share the same eigenvectors, and so do $\widehat{B}$ and $|\widehat{B}|$. With the linearization of the artificial dissipation included on the left-hand side, the diagonal form requires the solution of scalar pentadiagonal rather than block pentadiagonal systems, which results in a significant saving in computational work for the solution of steady flows.

The diagonal algorithm is an efficient and robust algorithm. However, there are some cases with specific properties for which it will not converge; in such cases, the block pentadiagonal algorithm is more reliable. An intermediate block form in which block tridiagonal systems are solved has also received considerable use. In this intermediate approach, the contribution of the fourth-difference dissipation on the left-hand side is approximated by a second-difference dissipation term with a coefficient equal to twice the coefficient of the fourth-difference dissipation on the right-hand side. It can be shown using linear theory that this approximation remains unconditionally stable. Such an algorithm will typically converge much more slowly than a full pentadiagonal linearization, but it has a lower cost per time step than the block pentadiagonal algorithm and can be more robust than the scalar pentadiagonal algorithm in some cases.

4.5.6 Convergence Acceleration for Steady Flow Computations

Local Time Stepping. As discussed in Sect. 4.5.4, the approximate factorization leads to an amplification factor σ that approaches unity as the time step h tends to infinity. Consequently, there is an optimum time step that minimizes the maximum magnitude of σ for the various eigenvalues associated with the Jacobian of the discrete spatial operator and hence produces the fastest possible convergence to steady state. For the inviscid flux terms, the eigenvalues of the Jacobian of the discrete residual vector are proportional to the characteristic speeds, e.g. u, $u+a$, $u-a$ in one dimension, divided by a characteristic mesh spacing, e.g. Δx in one dimension. The amplification factor σ is a function of the product of the eigenvalues and the time step h. Hence the convergence rate is dependent on the Courant (or CFL) number, given in one dimension by

$$C_{\mathrm{n}} = \frac{(|u| + a)h}{\Delta x}. \tag{4.138}$$

Here we have defined the Courant number based on the largest characteristic speed, $|u|+a$, but waves propagating at the other characteristic speeds will have a different effective Courant number.

Both the characteristic speeds and the mesh spacing can vary widely within a mesh. With a constant h, the local Courant number associated with each mesh node will thus also vary widely and will be suboptimal. When computing steady flows, we have the freedom to vary the time step locally in space. This destroys time accuracy but has no effect on the converged steady-state solution. Local time stepping can have a substantial influence on the convergence rate of a factored algorithm. It can be viewed as a way to condition the iteration matrix of the iterative methods defined via (4.118) or (4.128), or it can be interpreted as an attempt to use a more uniform (and hence closer to optimal) Courant number throughout the flow field. In any event, local time stepping can be effective for grid spacings that vary from very fine to very coarse—a situation usually encountered in simulations that contain a wide variety of length scales.

As a rule, one wishes to adjust the local time step at each grid node in proportion to the local grid spacing divided by the local characteristic speed of the flow, leading to a constant Courant number. In multiple dimensions, the situation is not quite so straightforward. For example, a cell with a high aspect ratio has two distinct grid spacings. In two dimensions, an approximation to a constant Courant number is achieved by the following formula for the local time step:

$$\Delta t = \frac{\Delta t_{\text{ref}}}{|U| + |V| + a\sqrt{\xi_x^2 + \xi_y^2 + \eta_x^2 + \eta_y^2}}, \tag{4.139}$$

where Δt_{ref} is defined by the user and must be chosen through experimentation to provide fast convergence.

For highly stretched grids, the grid spacing can vary by over six orders of magnitude. The variation in the characteristic speeds is generally more moderate. Therefore, the grid spacing is the more important parameter for maintaining a reasonably uniform Courant number, and a purely geometric variation of Δt can be effective. The following geometric formula for the local time step produces fast convergence when used with the approximately factored algorithm [2]:

$$\Delta t = \frac{\Delta t|_{\text{ref}}}{1 + \sqrt{J}}. \tag{4.140}$$

The term J^{-1} is closely related to the cell area. Therefore, this formula produces a Δt that is roughly proportional to the square root of the cell area. The addition of unity to the denominator prevents Δt from becoming too large at the largest grid cells.

To illustrate the advantage of using a variable time step, Fig. 4.5 shows the improvement in convergence rate when a variable time step based on (4.140) is substituted for a constant time step in a NACA 0012 airfoil test case where the Euler

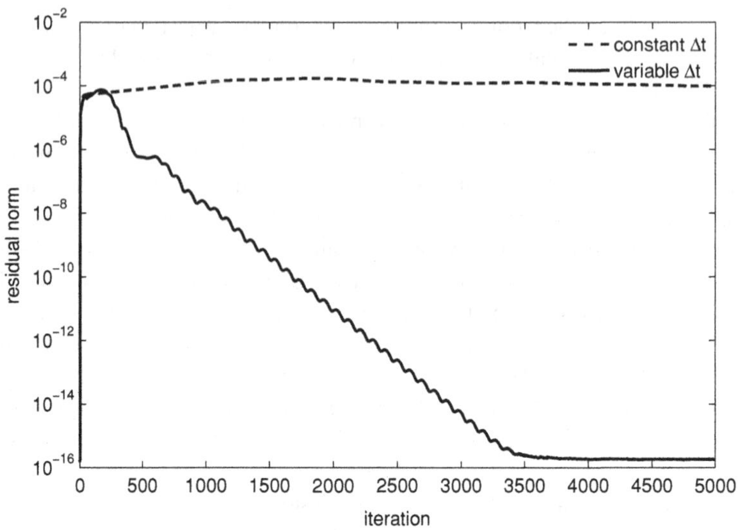

Fig. 4.5 Convergence improvement due to local time stepping

equations are solved at a Mach number of 0.8 and an angle of attack of 1.25 degrees. The constant time step chosen is the largest stable constant time step. For this comparison all other parameters were held constant.

In the above discussion, we have considered only the local Courant number, which is related to the inviscid fluxes. For an implicit algorithm, determination of the local time step based on inviscid considerations is generally sufficient for high Reynolds number flows, as these are convection dominated. For flows at low Reynolds numbers, consideration also needs to be given to the local Von Neumann number (see Sect. 2.7.4). As we will see in Chap. 5, local time stepping is even more critical for explicit methods.

Mesh Sequencing. The mesh density is based on accuracy considerations. A sufficiently fine mesh must be used such that the numerical errors from the spatial discretization lie below a desired threshold. The iterative methods given by (4.118) or (4.128) require an initial solution to begin the process. Fewer iterations are needed to converge to steady state if the initial solution is not too far from the converged solution, which is of course unknown at the outset. It is common to initiate the iterations with a solution given by a uniform flow that satisfies some free-stream or inflow boundary conditions. This provides a relatively poor initial guess that is much different from the eventual steady solution. Therefore, one way to improve convergence is to begin the iterations using a much coarser mesh than that dictated by the accuracy requirements. On a coarse mesh, the iterations will converge with relatively little computational work to a solution that provides a much improved initial guess for the fine mesh iterations. The solution obtained after reducing the norm of the residual on the coarse mesh by a few orders of magnitude can be interpolated onto

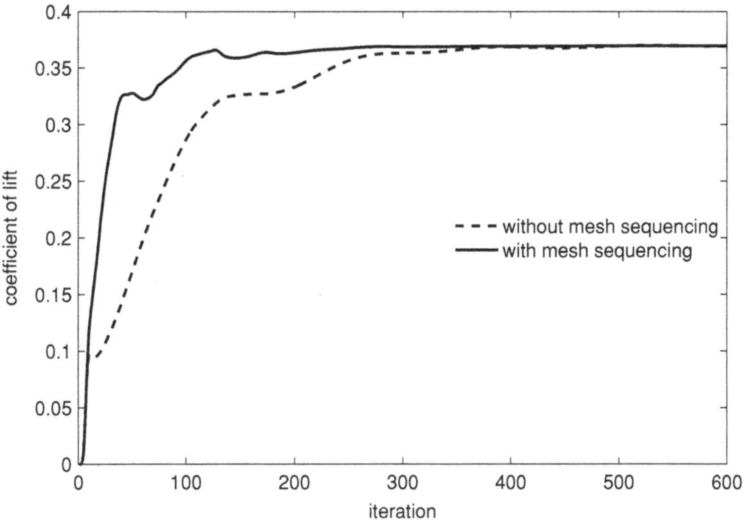

Fig. 4.6 Improvement in convergence of lift coefficient due to mesh sequencing

the finer mesh to provide the initial iterate for the iterations on the fine mesh. This process can be repeated on a sequence of meshes, beginning with a very coarse mesh and ending on the fine mesh dictated by accuracy requirements. The use of mesh sequencing in this manner can also improve the robustness of a solver, as the coarse meshes are effective at damping initial transients, when nonlinear effects are large.

Figure 4.6 shows an example of the improvement in convergence resulting from mesh sequencing. For an inviscid flow over the NACA 0012 airfoil at a Mach number of 0.8 and an angle of attack of 1.25 degrees, a sequence of four C-meshes has been used. The first mesh is 32 by 17, the second 63 by 33, the third 125 by 69, and the final mesh has 249 by 98 nodes. Both cases were started with a free-stream initial condition.

4.5.7 Dual Time Stepping for Unsteady Flow Computations

The implicit algorithm described above is suitable for time-accurate computations of unsteady flows where the equations are integrated through time from some meaningful initial condition. A sufficiently fine mesh is needed to ensure that spatial discretization errors are small; in addition, the time step must be selected such that the temporal discretization errors are also below the desired threshold. Generally speaking, at least second-order temporal accuracy is desired. The local time linearization and approximate factorization preserve the order of accuracy of a second-order implicit time-marching method, such as the second-order backward and trapezoidal methods discussed earlier. Neither the diagonal form nor local time stepping should be used for time-accurate computations of unsteady flows.

The second-order backwards time-marching method is given by

$$u_{n+1} = \frac{1}{3}[4u_n - u_{n-1} + 2hu'_{n+1}]. \tag{4.141}$$

Applying this method to the thin-layer form of (4.95) gives

$$\begin{aligned}\widehat{Q}^{n+1} &= \frac{4}{3}\widehat{Q}^n - \frac{1}{3}\widehat{Q}^{n-1}\\ &+\frac{2h}{3}\left(-\delta_\xi \widehat{E}^{n+1} + D_\xi^{n+1} - \delta_\eta \widehat{F}^{n+1} + D_\eta^{n+1} + Re^{-1}\delta_\eta \widehat{S}^{n+1}\right).\end{aligned} \tag{4.142}$$

After local time linearization and approximate factorization, a form analogous to (4.118) is obtained:

$$\begin{aligned}&\left[I + \frac{2h}{3}\delta_\xi \widehat{A}^n\right]\left[I + \frac{2h}{3}\delta_\eta \widehat{B}^n - \frac{2h}{3}Re^{-1}\delta_\eta \widehat{M}^n\right]\Delta\widehat{Q}^n\\ &= \widehat{Q}^n - \widehat{Q}^{n-1} - \frac{2h}{3}\left[\delta_\xi \widehat{E}^n + \delta_\eta \widehat{F}^n - Re^{-1}\delta_\eta \widehat{S}^n\right].\end{aligned} \tag{4.143}$$

The method given by (4.143) is the approximately factored form of the second-order backward time-marching method. It is an efficient second-order implicit method for time-accurate computations of unsteady flows. However, despite the fact that the linearization and factorization errors do not diminish the order of accuracy of the method, they increase the error incurred per time step. This is the motivation for the dual time stepping approach, which eliminates linearization and factorization errors.

In order to demonstrate the dual time stepping approach, we begin by rearranging (4.142) as follows

$$\frac{3\widehat{Q}^{n+1} - 4\widehat{Q}^n + \widehat{Q}^{n-1}}{2h} + R(\widehat{Q}^{n+1}) = 0, \tag{4.144}$$

where

$$R(\widehat{Q}^{n+1}) = \left[\delta_\xi \widehat{E}^{n+1} - D_\xi^{n+1} + \delta_\eta \widehat{F}^{n+1} - D_\eta^{n+1} - Re^{-1}\delta_\eta \widehat{S}^{n+1}\right]. \tag{4.145}$$

This is a nonlinear algebraic equation that must be solved for $\widehat{Q}^{n+1}$ at each time step. To reflect this, we define $R_u(\widehat{Q})$ as

$$R_u(\widehat{Q}) = \frac{3\widehat{Q} - 4\widehat{Q}^n + \widehat{Q}^{n-1}}{2h} + R(\widehat{Q}), \tag{4.146}$$

so the nonlinear equation to be solved is simply

$$R_{\mathrm{u}}(\widehat{Q}) = 0. \tag{4.147}$$

One can readily observe the similarity between the nonlinear equation to be solved at every iteration of the second-order backward time-marching method, $R_{\mathrm{u}}(\widehat{Q}) = 0$, and the equation to be solved for a steady flow, $R(\widehat{Q}) = 0$. Therefore, any method developed for steady problems, such as inexact-Newton methods and implicit or explicit time-marching methods that follow a time-dependent path to steady state, can be used to solve (4.147).

In this chapter, our focus is on the approximate factorization algorithm, which follows a time-dependent, though not necessarily time-accurate, path to the steady solution. In order to enable application of this algorithm to the solution of (4.147), we introduce a pseudo-time variable τ (not to be confused with the variable τ in the generalized curvilinear coordinate transformation) to produce a system of ODEs as follows:

$$\frac{\mathrm{d}\widehat{Q}}{\mathrm{d}\tau} + R_{\mathrm{u}}(\widehat{Q}) = 0. \tag{4.148}$$

In order to solve for the steady-state solution of this ODE, which is the solution to (4.147), we can apply the approximately-factored implicit Euler method. We introduce a pseudo-time index p such that $\widehat{Q}^p = \widehat{Q}(p\Delta\tau)$, where $\Delta\tau = \tau_{p+1} - \tau_p$, to obtain

$$\begin{aligned}&\left[I + \frac{\Delta\tau}{b}\delta_\xi \widehat{A}^p\right]\left[I + \frac{\Delta\tau}{b}\delta_\eta \widehat{B}^p - \frac{\Delta\tau}{b} Re^{-1}\delta_\eta \widehat{M}^p\right]\Delta\widehat{Q}^p \\ &= -\frac{\Delta\tau}{b} R_{\mathrm{u}}(\widehat{Q}^p),\end{aligned} \tag{4.149}$$

where

$$b = 1 + \frac{3\Delta\tau}{2h},$$

and we have divided by b before factoring. The converged solution obtained from this iterative process provides $\widehat{Q}^{n+1}$. The accuracy of the time-marching method is dictated by the time step h, while the pseudo-time step $\Delta\tau$ can be chosen for fast convergence with no regard for time accuracy, since it has no effect on the converged solution of (4.147). Similarly, for the pseudo-time iterations, the diagonal form and local time stepping can be used to speed up convergence.

Dual time stepping is an example of an approach where an iterative method is used to solve the nonlinear equation that arises at each time step of an implicit method. This approach eliminates linearization and factorization errors and can also simplify the implementation of boundary conditions. It is natural to use a fast steady solver for the solution of this nonlinear equation along with any convergence acceleration techniques developed for steady flows. One may question the efficiency of an approach where the unsteady problem is in effect solved as a sequence of steady problems. However, it is important to note that the initial iterate for the pseudo-time iterations is $\widehat{Q}^n$, which is a much better estimate of $\widehat{Q}^{n+1}$ then is usually available for steady computations. Hence one can expect the number pseudo-time steps needed to obtain a converged solution to (4.147) to be much less than the number of time steps needed to obtain a converged solution to a steady flow problem.

4.6 Boundary Conditions

There are a number of different ways to implement boundary conditions. Before describing one particular approach, we will introduce the important aspects of boundary condition development that must be considered in selecting an approach, which are as follows:

1. The physical definition of the flow problem must be properly represented. For example, a viscous flow ordinarily requires a no-slip condition at solid surfaces.
2. The physical conditions must be expressed in mathematical form and must be consistent with the mathematical description of the problem. For example, the no-slip condition referred to above must be expressed in terms of the variables selected. Moreover, this condition cannot be applied if inviscid governing equations are chosen.
3. The boundary conditions expressed in mathematical form must be approximated numerically.
4. Depending on the algorithm, the numerical scheme in the interior may require more boundary information than the physics provides. Hence a means must be developed for providing this additional boundary information.
5. The combination of the interior scheme with the boundary scheme must be checked for stability and accuracy. In general, the two should have consistent accuracy.
6. The boundary condition formulation must be assessed in terms of its impact on the efficiency and generality of the solver.

With these considerations in mind, one can approach the development of boundary conditions from several different directions. Moreover, there exist various different boundary types, such as inflow/outflow boundaries, solid walls, symmetry boundaries, and periodic boundaries, one or more of which can be present in a specific flow problem. In this chapter, we will cover an approach to the boundary conditions typically associated with computations of external flows. The basic principles covered are easily extended to other boundary types.

With an implicit solver, one might expect implicit boundary conditions to be a strict requirement. In order to obtain the benefits of an inexact-Newton method, they are certainly recommended. For an approximately-factored solver, however, the optimal time step is not so large that implicit boundary conditions are essential, and the use of explicit boundary conditions does not typically degrade the convergence rate.

For external flows, one is faced with the problem that the boundary conditions are defined at an infinite distance from the body. Although coordinate transformations can be used to address this, it is much more common to introduce an artificial far-field boundary in order to limit the size of the computational domain. This boundary must be located a sufficient distance from the body that the error introduced does not exceed the desired error threshold. At a far-field boundary, viscous effects are typically negligible and the flow can be considered inviscid. Consequently, a characteristic approach is taken to inflow and outflow boundary conditions at the far-field boundary. Proper application of characteristic theory is essential in order to ensure well-posedness. At a far-field boundary through which a wake is advecting or viscous effects are not negligible, a different approach is used; this is discussed further below.

4.6.1 Characteristic Approach

The concept of characteristic theory is most easily demonstrated with the linearized one-dimensional Euler equations, where

$$\partial_t Q + \partial_x (AQ) = 0 \tag{4.150}$$

represents the model equation. Since A is a constant-coefficient matrix, we can diagonalize (4.150) using the relation $A = X\Lambda_A X^{-1}$, where X is the right eigenvector matrix, and

$$\Lambda_A = \begin{bmatrix} u & 0 & 0 \\ 0 & u+a & 0 \\ 0 & 0 & u-a \end{bmatrix}. \tag{4.151}$$

Premultiplying by X^{-1} and inserting the product XX^{-1} after A, we obtain

$$\partial_t \left(X^{-1}Q\right) + \Lambda_A \partial_x \left(X^{-1}Q\right) = 0. \tag{4.152}$$

Defining $X^{-1}Q = W$, we now have a diagonal system. The equations have been decoupled into three equations in the form of the linear convection equation with the characteristic speeds u, $u+a$, and $u-a$. The associated characteristic variables, or Riemann invariants, for this constant-coefficient linear system are defined by W. One can also obtain these same characteristic speeds and the associated Riemann invariants for the full nonlinear Euler equations without the assumption that A is a constant coefficient matrix.

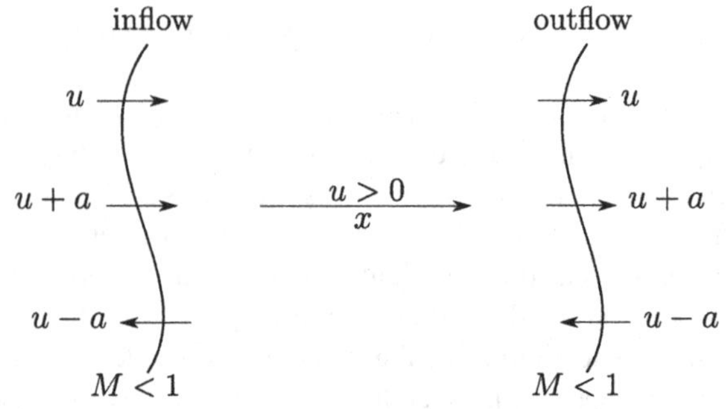

Fig. 4.7 Characteristics at subsonic inflow and outflow boundaries of a closed domain

With the diagonalized form of the equations, the boundary condition requirements are clear. Consider first a subsonic flow. At the left boundary of a closed physical domain, see Fig. 4.7, where $0 < u < a$ (for example, subsonic inflow for a channel flow), the two characteristic speeds $u, u + a$ are positive, while $u - a$ is negative. At inflow then, two pieces of information enter the domain along the two incoming characteristics, and one piece leaves along the outgoing characteristic. At the outflow boundary, one piece of information enters and two leave. Thus we can obtain a well-posed problem by specifying the first two components of W, which are the two incoming characteristic variables, at the inflow boundary and then handling the third characteristic variable such that its value is not constrained, i.e. it is determined by the interior flow. At the outflow boundary, we specify the third component of W and determine the first two from the interior flow. If the flow is supersonic, all characteristic speeds have the same sign. Hence one must specify all variables at inflow and none at outflow.

It is not necessary to specify the characteristic variables; other flow quantities can be used, as long as they lead to well-posed conditions. The major constraint is that the correct number of boundary values corresponding to incoming characteristics must be specified, regardless of the variables that are chosen. Some combinations of variables lead to a well-posed problems, others do not. In the next section, we describe a test to establish whether a given choice of variables is well posed.

4.6.2 Well-Posedness Test

A check on the well posedness of boundary conditions is given by Chakravarthy [17]. Let us consider one-dimensional flow with subsonic inflow and subsonic outflow. Then two variables can be specified at inflow, associated with the first two eigenvalues, and one variable can be specified at outflow, associated with the third eigenvalue. As an example, we test the following specified values: $\rho = \rho_{\text{in}}$, $\rho u = (\rho u)_{\text{in}}$ and $p = p_{\text{out}}$. These can be written as

$$B_{\text{in}}(Q) = \begin{bmatrix} q_1 \\ q_2 \\ 0 \end{bmatrix} = B_{\text{in}}(Q_{\text{in}}), \tag{4.153}$$

$$B_{\text{out}}(Q) = \begin{bmatrix} 0 \\ 0 \\ (\gamma - 1)(q_3 - \frac{1}{2}q_2^2/q_1) \end{bmatrix} = B_{\text{out}}(Q_{\text{out}}), \tag{4.154}$$

with $q_1 = \rho,\ q_2 = \rho u,\ q_3 = e$.

Forming the Jacobians $C_{\text{in}} = \partial B_{\text{in}}/\partial Q$, and $C_{\text{out}} = \partial B_{\text{out}}/\partial Q$ we have

$$C_{\text{in}} = \begin{bmatrix} 1\,0\,0 \\ 0\,1\,0 \\ 0\,0\,0 \end{bmatrix}, \quad C_{\text{out}} = \begin{bmatrix} 0 & 0 & 0 \\ 0 & 0 & 0 \\ ((\gamma-1)/2)\,u^2 & -(\gamma-1)u & \gamma-1 \end{bmatrix}. \tag{4.155}$$

The left eigenvector matrix X^{-1} for the one-dimensional Euler equations is[3]

$$\begin{bmatrix} 1 - \frac{u^2}{2}(\gamma-1)a^{-2} & (\gamma-1)ua^{-2} & -(\gamma-1)a^{-2} \\ \beta[(\gamma-1)\frac{u^2}{2} - ua] & \beta[a - (\gamma-1)u] & \beta(\gamma-1) \\ \beta[(\gamma-1)\frac{u^2}{2} + ua] & -\beta[a + (\gamma-1)u] & \beta(\gamma-1) \end{bmatrix}, \tag{4.156}$$

with $\beta = 1/(\sqrt{2}\rho a)$.

The condition for well-posedness of these example boundary conditions is that $\overline{C}_{\text{in}}^{-1}$ and $\overline{C}_{\text{out}}^{-1}$ exist, where

$$\overline{C}_{\text{in}} = \begin{bmatrix} 1 & 0 & 0 \\ 0 & 1 & 0 \\ \beta[(\gamma-1)\frac{u^2}{2} + ua] & -\beta[a + (\gamma-1)u] & \beta(\gamma-1) \end{bmatrix}, \tag{4.157}$$

and

$$\overline{C}_{\text{out}} = \begin{bmatrix} 1 - \frac{u^2}{2}(\gamma-1)a^{-2} & (\gamma-1)ua^{-2} & -(\gamma-1)a^{-2} \\ \beta[(\gamma-1)\frac{u^2}{2} - ua] & \beta[a - (\gamma-1)u] & \beta(\gamma-1) \\ (\gamma-1)\frac{u^2}{2} & -(\gamma-1)u & \gamma-1 \end{bmatrix}. \tag{4.158}$$

These matrices are formed by adjoining the eigenvectors associated with the outgoing characteristics at the boundary in question to the Jacobian matrices of the boundary conditions. The inverses of the above matrices will exist if their determinants are nonzero. For the two boundaries, we have $\det(\overline{C}_{\text{in}}) = \beta(\gamma-1) \neq 0$, and $\det(\overline{C}_{\text{out}}) = \beta(\gamma-1)a \neq 0$. Therefore, this particular choice of boundary conditions is well posed. Other choices for specified boundary values can be similarly checked.

---

[3] The rows of X^{-1} are the left eigenvectors of A.

4.6.3 Boundary Conditions for External Flows

We shall outline below some of the more commonly used boundary conditions. These will be presented in the context of a body-fitted C mesh, as depicted in Fig. 4.2, and are easily generalized to other mesh topologies. The approach taken is to solve the governing equations only at the interior nodes of the mesh. Therefore, all variables must be given at the boundary by the numerical boundary conditions. Since the physical boundary conditions provide boundary values for only some of the variables, the others must be determined by extrapolation from the interior flow solution. Moreover, the numerical boundary conditions can be implemented either explicitly or implicitly. In an explicit treatment, the boundary values are held fixed during one iteration of the approximate factorization algorithm. They are then updated based on the new $\widehat{Q}$, and the process is repeated. For an implicit implementation, the numerical boundary conditions must be linearized and the appropriate terms included in the left-hand-side operator of the implicit algorithm.

Body Surfaces. At a body surface, tangency must be satisfied for inviscid flow and the no-slip condition for viscous flow. In two-dimensions, body surfaces are usually mapped to η = constant coordinates, as in Fig. 4.2. In this case, as shown in Sect. 4.2.4, the normal component of velocity is given in terms of the metrics of the transformation by

$$V_n = \frac{\eta_x u + \eta_y v}{\sqrt{\eta_x^2 + \eta_y^2}}, \tag{4.159}$$

and the tangential component by

$$V_t = \frac{\eta_y u - \eta_x v}{\sqrt{\eta_x^2 + \eta_y^2}}. \tag{4.160}$$

For inviscid flows, flow tangency is satisfied by setting $V_n = 0$. The tangential velocity V_t is obtained at the body surface through linear extrapolation along the coordinate line approaching the surface, using the interior values of Q at the nodes above the surface. It is preferable to extrapolate Cartesian velocity components and then form the tangential velocity component based on the extrapolated values. The Cartesian velocity components at the surface are found from the following relation obtained by solving (4.159) and (4.160) for u and v:

$$\begin{pmatrix} u \\ v \end{pmatrix} = \frac{1}{\sqrt{\eta_x^2 + \eta_y^2}} \begin{bmatrix} \eta_y & \eta_x \\ -\eta_x & \eta_y \end{bmatrix} \begin{pmatrix} V_t \\ V_n \end{pmatrix}, \tag{4.161}$$

with V_n set to zero, and V_t determined from the extrapolation. For a viscous flow, the no-slip condition gives $u = v = 0$.

For an inviscid flow, flow tangency is the only physical boundary condition. Therefore only one variable can be specified, which is the normal velocity component, and three more variables must be determined from the interior flow solution. The tangential velocity component is extrapolated, as described above. In addition, pressure and density, for example, can be extrapolated. For steady inviscid flows with uniform upstream conditions, the total or stagnation enthalpy ($H = (e + p)/\rho$) is constant, at least in the exact solution. This requirement can be exploited to determine one variable. For example, after u, v, and p are obtained at the surface, the density can be found by requiring that the total enthalpy at the boundary be equal to the free-stream total enthalpy. Once boundary values for u, v, p, and ρ are determined, the corresponding conservative variables are easily found using their definitions along with the equation of state.

For viscous flows, there is an additional boundary condition related to heat transfer that determines the temperature or its gradient normal to the surface. If the wall remains at constant temperature, then this temperature must be specified. More commonly, an adiabatic condition is appropriate. In this case, there is no heat transfer to or from the wall, giving

$$\frac{\partial T}{\partial n} = 0, \tag{4.162}$$

where n is the direction normal to the wall, and the derivative must be approximated numerically using a one-sided difference formula. This condition provides the temperature at the wall. The wall pressure can be determined by extrapolation from the interior; the conservative variables can then be found from the values of u, v, T, and p.

Far-Field Boundaries. The far-field boundary must be located a sufficient distance away from the body that its effect on the computed solution is negligible. This can be determined by experimentation. The basic goal of the boundary conditions at the far-field boundary is to permit disturbances to exit the domain with little or no reflection, as such artificial reflections can pollute the solution in the interior of the domain. For problems where accurate propagation of waves to and through the outer boundary is critical, specialized non-reflecting boundary conditions have been developed (see for example the discussion by Colonius and Lele [18]). For many flow problems, non-reflecting boundary conditions based on the method of characteristics are sufficient; these are described here.

Following the discussion in Sect. 4.6.1, the idea is to specify incoming Riemann invariants and determine outgoing Riemann invariants from the interior solution by extrapolation. For subsonic flows, we describe an extension to two dimensions based on locally one-dimensional Riemann invariants. The relevant velocity component is that normal to the outer boundary V_n. With n pointing outward from the flow domain, a positive V_n defines an outflow boundary, while a negative V_n defines an inflow boundary. As shown in the Appendix, the two-dimensional inviscid flux Jacobians have three distinct eigenvalues, with the eigenvalue corresponding to the convective speed repeated. From the one-dimensional theory, we have three Riemann invariants, so one more variable is needed in two dimensions that will be associated with the repeated eigenvalue. The velocity component tangential to the boundary can be used for this purpose. Therefore, we have the following characteristic speeds and associated variables:

$$\begin{aligned}
\lambda_1 &= V_n - a, \quad & R_1 &= V_n - 2a/(\gamma - 1) \\
\lambda_2 &= V_n + a, \quad & R_2 &= V_n + 2a/(\gamma - 1) \\
\lambda_3 &= V_n, \quad & R_3 &= S = \ln \frac{p}{\rho^\gamma} \quad \text{(entropy)} \\
\lambda_4 &= V_n, \quad & R_4 &= V_t.
\end{aligned} \tag{4.163}$$

For a subsonic inflow boundary, where $V_n < 0$, the characteristic speeds satisfy the following:

$$\lambda_1 < 0, \ \lambda_2 > 0, \ \lambda_3 < 0, \ \lambda_4 < 0.$$

A negative characteristic speed corresponds to an incoming characteristic; hence the associated variables must be specified based on free-stream values. The variables associated with positive characteristic speeds must be determined from the interior flow. In this case, R_1, R_3, and R_4 must be specified, and R_2 must be extrapolated from the interior. Once these four variables are determined at the boundary, the four conservative variables can be obtained.

For a subsonic outflow boundary, where $V_n > 0$, the eigenvalues satisfy the following:

$$\lambda_1 < 0, \ \lambda_2 > 0, \ \lambda_3 > 0, \ \lambda_4 > 0.$$

Therefore, R_1 must be set to its free-stream value, and R_2, R_3, and R_4 must be extrapolated from the interior.

For supersonic inflow boundaries, all flow variables are specified; for supersonic outflow boundaries, all variables are extrapolated. For a subsonic boundary through which a viscous wake is flowing, all variables are extrapolated (see [19] for a detailed discussion). Special treatments may be needed at interfaces between blocks in multi-block meshes or at wake cuts. See, for example, Osusky and Zingg [20].

Far-Field Circulation Correction. For computations of two-dimensional flows over lifting bodies, the far-field circulation correction reduces the effect of the far-field boundary location. This enables the distance to the far-field boundary to be reduced without compromising accuracy.Far from a lifting airfoil in a subsonic free-stream, the perturbation caused by the airfoil approaches that induced by a point vortex. This can be exploited by adding the perturbation associated with a point vortex to the free-stream values when applying the far-field boundary conditions.

Following Salas et al. [21], a compressible potential vortex solution is added as a perturbation to the free-stream quantities at the far-field boundary. With the present nondimensionalization, the free-stream velocity components are $u_\infty = M_\infty \cos\alpha$

and $v_\infty = M_\infty \sin\alpha$, where M_∞ is the free-stream Mach number, and α is the angle of incidence of the flow relative to the x axis. The perturbed far-field boundary velocities are defined as

$$u_{\mathrm{f}} = u_\infty + \frac{\beta \Gamma \sin(\theta)}{2\pi r \left(1 - M_\infty^2 \sin^2(\theta - \alpha)\right)} \tag{4.164}$$

and

$$v_{\mathrm{f}} = v_\infty - \frac{\beta \Gamma \cos(\theta)}{2\pi r \left(1 - M_\infty^2 \sin^2(\theta - \alpha)\right)}, \tag{4.165}$$

where the circulation $\Gamma = \frac{1}{2} M_\infty l C_l$, l is the chord length, C_l is the coefficient of lift, α is the angle of attack, $\beta = \sqrt{1 - M_\infty^2}$, and r, θ are polar coordinates to the point of application on the outer boundary relative to an origin at the quarter-chord point on the airfoil center line. A corrected speed of sound is used that enforces constant free-stream enthalpy at the boundary:

$$a_{\mathrm{f}}^2 = (\gamma - 1)\left(H_\infty - \frac{1}{2}(u_{\mathrm{f}}^2 + v_{\mathrm{f}}^2)\right). \tag{4.166}$$

Equations (4.164), (4.165) and (4.166) are used instead of free-stream values in defining the specified quantities for the far-field characteristic boundary conditions. The circulation Γ is determined by the solution and is not known at the outset; hence it must be calculated and updated as the iterations progress. At convergence, the value of Γ used in the far-field circulation correction is consistent with the lift coefficient computed for the airfoil.

Figure 4.8 shows the coefficient of lift C_l plotted against the inverse of the distance to the outer boundary for an inviscid flow over the NACA 0012 airfoil at $M_\infty = 0.63$, $\alpha = 2.0$ degrees. The distance to the outer boundary varies from 5 to 200 chord lengths, where outer mesh rings were eliminated from the largest mesh to produce the smaller meshes.

4.7 Three-Dimensional Algorithm

The three-dimensional form of the implicit algorithm follows the same development as the two-dimensional algorithm. The curvilinear coordinate transformation is carried out in the same fashion. The block and diagonal algorithms take the same format. Boundary conditions are analogous. In this section, we briefly outline the equations in three dimensions.

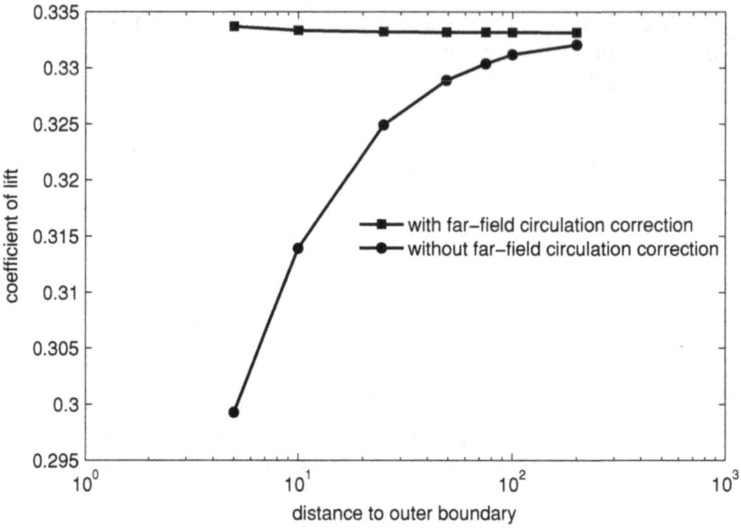

Fig. 4.8 Effect on lift coefficient of varying outer boundary distance (in chord lengths) with and without far-field circulation correction

4.7.1 Flow Equations

The full three-dimensional Navier-Stokes equations in strong conservation law form are reduced to the thin-layer form under the same restrictions and assumptions as in two dimensions. The equations in generalized curvilinear coordinates are

$$\partial_\tau \widehat{Q} + \partial_\xi \widehat{E} + \partial_\eta \widehat{F} + \partial_\zeta \widehat{G} = Re^{-1} \partial_\zeta \widehat{S}, \tag{4.167}$$

where

$$\widehat{Q} = J^{-1} \begin{bmatrix} \rho \\ \rho u \\ \rho v \\ \rho w \\ e \end{bmatrix}, \quad \widehat{E} = J^{-1} \begin{bmatrix} \rho U \\ \rho u U + \xi_x p \\ \rho v U + \xi_y p \\ \rho w U + \xi_z p \\ U(e+p) - \xi_t p \end{bmatrix},$$

$$\widehat{F} = J^{-1} \begin{bmatrix} \rho V \\ \rho u V + \eta_x p \\ \rho v V + \eta_y p \\ \rho w V + \eta_z p \\ V(e+p) - \eta_t p \end{bmatrix}, \quad \widehat{G} = J^{-1} \begin{bmatrix} \rho W \\ \rho u W + \zeta_x p \\ \rho v W + \zeta_y p \\ \rho w W + \zeta_z p \\ W(e+p) - \zeta_t p \end{bmatrix}, \tag{4.168}$$

with

$$
\begin{aligned}
U &= \xi_t + \xi_x u + \xi_y v + \xi_z w, \\
V &= \eta_t + \eta_x u + \eta_y v + \eta_z w \\
W &= \zeta_t + \zeta_x u + \zeta_y v + \zeta_z w,
\end{aligned} \tag{4.169}
$$

and

$$
\widehat{S} = J^{-1} \begin{bmatrix} 0 \\ \mu m_1 u_\zeta + (\mu/3) m_2 \zeta_x \\ \mu m_1 v_\zeta + (\mu/3) m_2 \zeta_y \\ \mu m_1 w_\zeta + (\mu/3) m_2 \zeta_z \\ \mu m_1 m_3 + (\mu/3) m_2 (\zeta_x u + \zeta_y v + \zeta_z w) \end{bmatrix}. \tag{4.170}
$$

Here $m_1 = \zeta_x^2 + \zeta_y^2 + \zeta_z^2$, $m_2 = \zeta_x u_\zeta + \zeta_y v_\zeta + \zeta_z w_\zeta$, and $m_3 = (u^2 + v^2 + w^2)_\zeta/2 + Pr^{-1}(\gamma - 1)^{-1}(a^2)_\zeta$. Pressure is again related to the conservative flow variables, Q, by the equation of state:

$$
p = (\gamma - 1)\left(e - \frac{1}{2}\rho(u^2 + v^2 + w^2)\right). \tag{4.171}
$$

The metric terms are defined as

$$
\begin{aligned}
\xi_x &= J(y_\eta z_\zeta - y_\zeta z_\eta), & \eta_x &= J(z_\xi y_\zeta - y_\xi z_\zeta) \\
\xi_y &= J(z_\eta x_\zeta - z_\zeta x_\eta), & \eta_y &= J(x_\xi z_\zeta - z_\xi x_\zeta) \\
\xi_z &= J(x_\eta y_\zeta - y_\eta x_\zeta), & \eta_z &= J(y_\xi x_\zeta - x_\xi y_\zeta) \\
\zeta_x &= J(y_\xi z_\eta - z_\xi y_\eta), & \xi_t &= -x_\tau \xi_x - y_\tau \xi_y - z_\tau \xi_z \\
\zeta_y &= J(z_\xi x_\eta - x_\xi z_\eta), & \eta_t &= -x_\tau \eta_x - y_\tau \eta_y - z_\tau \eta_z \\
\zeta_z &= J(x_\xi y_\eta - y_\xi x_\eta), & \zeta_t &= -x_\tau \zeta_x - y_\tau \zeta_y - z_\tau \zeta_z
\end{aligned} \tag{4.172}
$$

with

$$
J^{-1} = x_\xi y_\eta z_\zeta + x_\zeta y_\xi z_\eta + x_\eta y_\zeta z_\xi - x_\xi y_\zeta z_\eta - x_\eta y_\xi z_\zeta - x_\zeta y_\eta z_\xi. \tag{4.173}
$$

4.7.2 Numerical Methods

The implicit approximate factorization algorithm applied to the three-dimensional equations is

$$
\begin{aligned}
&\left[I + h\delta_\xi \widehat{A}^n\right]\left[I + h\delta_\eta \widehat{B}^n\right]\left[I + h\delta_\zeta \widehat{C}^n - hRe^{-1}\delta_\zeta \widehat{M}^n\right] \Delta\widehat{Q}^n \\
&= -h\left(\delta_\xi \widehat{E}^n + \delta_\eta \widehat{F}^n + \delta_\zeta \widehat{G}^n - Re^{-1}\delta_\zeta \widehat{S}^n\right).
\end{aligned} \tag{4.174}
$$

The three-dimensional inviscid flux Jacobians $\widehat{A}, \widehat{B}, \widehat{C}$ are defined in the Appendix along with the viscous flux Jacobian $\widehat{M}$. The spatial discretization, including the artificial dissipation, extends directly to three dimensions. Calculation of the grid metrics in three dimensions is discussed in Sect. 4.4.1. The diagonal algorithm in three dimensions has the form

$$T_\xi \left[I + h\,\delta_\xi\,\Lambda_\xi\right] \widehat{N} \left[I + h\;\delta_\eta\,\Lambda_\eta\right] \widehat{P} \left[I + h\,\delta_\zeta\,\Lambda_\zeta\right] T_\zeta^{-1} \Delta \widehat{Q}^n = \widehat{R}^n \quad (4.175)$$

with $\widehat{N} = T_\xi^{-1} T_\eta$ and $\widehat{P} = T_\eta^{-1} T_\zeta$.

A linear constant-coefficient Fourier analysis for the three-dimensional model wave equation shows unconditional instability for the three-dimensional factored algorithm in the absence of numerical dissipation. This is due to the cross term errors. In contrast to the case of two dimensions where the cross term errors just affect the rapid convergence capability of the algorithm at large time steps, in three dimensions they result in a weak instability. The method becomes stable when a small amount of artificial dissipation is added to the spatial discretization.

4.8 One-Dimensional Examples

In order to demonstrate the performance of the algorithm presented in this chapter, we present numerical results obtained for steady flows governed by the quasi-one-dimensional Euler equations and an unsteady flow in a shock tube. The flow conditions coincide with those associated with the exercises of Chap. 3 and the present chapter. Hence the results presented in this section provide a useful reference for the reader when developing the code associated with this chapter's exercises. These one-dimensional problems should not be used to assess the efficiency of the algorithm, as their properties are simply too different from multi-dimensional problems. In particular, the implicit operator is tightly banded, which is not the case in multidimensions.

Three problems are considered, a subsonic channel flow, a transonic channel flow, and a shock tube. Flow conditions are as described in Sect. 3.3. The implicit algorithm is implemented as described in this chapter, although the coordinate transformation, the approximate factorization, and the viscous terms are not needed in this context. Boundary conditions are handled explicitly based on prescribing or extrapolating Riemann invariants. Zeroth-order extrapolation is used for outgoing Riemann invariants, i.e. the boundary value is set to the value at the first interior node. This is not desirable but leads to fast convergence for the two steady problems and has no impact on the shock-tube problem. Linear extrapolation is preferred and is needed to obtain second-order accuracy. It can be implemented through some minor changes to how the boundary values are handled (for example by choosing an updated boundary value that is the average of the value calculated using linear extrapolation and the previous value) or through an implicit treatment of the boundary conditions.

Alternatively, convergence can be obtained with linear extrapolation through the use of a low Courant number (e.g. $C_\mathrm{n} = 2$). In multidimensional external flows, good convergence can typically be obtained with linear extrapolation. Finally, in the implementation of the diagonal form, the contribution of the source term to the left-hand side operator is neglected.

The artificial dissipation coefficient values are $\kappa_2 = 0$, $\kappa_4 = 0.02$ for the subsonic channel flow problem, $\kappa_2 = 0.5$, $\kappa_4 = 0.02$ for both the transonic channel flow problem and the shock-tube problem. A nonzero value of κ_2 can be used for the subsonic problem but is not needed. The state at the inflow boundary is used as the initial condition for the channel flow problems. For these problems, which are steady, a local time step is calculated from (4.138) based on an input value of the Courant number. For the shock-tube problem, a constant time step is used based on an input Courant number and representative values of u and a. The values used are $u = 300\,\mathrm{m/s}$ and $a = 315\,\mathrm{m/s}$.

For the subsonic channel flow, Fig. 4.9 shows that the solution computed on a mesh with 49 interior nodes lies very close to the exact solution. Some oscillations are visible near the boundaries; these are associated with the zeroth-order extrapolation of the outgoing Riemann invariants. With linear extrapolation these are not seen. Results with 199 interior nodes are shown in Fig. 4.10; the oscillations are reduced.

One can compute the numerical error in density, for example, as

$$e_\rho = \sqrt{\sum_{j=1}^{M} \frac{(\rho_j - \rho_j^{\mathrm{exact}})^2}{M}}, \tag{4.176}$$

where M is the number of grid nodes, and ρ^{exact} is the exact solution. The error in density is plotted versus the grid spacing in Fig. 4.11. The numerical solution was obtained with linear extrapolation of the outgoing Riemann invariants at the boundaries and $\kappa_2 = 0$. The slope of the log-log plot is very close to two, consistent with second-order accuracy. This is a good test to verify a code.

Figures 4.12 and 4.13 display some convergence histories for the block form of the implicit algorithm applied to the subsonic channel problem. The L_2 norm of the residual is plotted versus the number of iterations for various grid sizes and Courant numbers. Figure 4.12 shows the dependence on the Courant number for a grid with 99 interior nodes, while Fig. 4.13 shows the dependence on the number of nodes in the grid with $C_\mathrm{n} = 40$.

The convergence of the diagonal form of the implicit algorithm is displayed in Fig. 4.14. The convergence behaviour of the diagonal form is comparable to that of the block form shown in Fig. 4.13. As a result, the savings associated with solving scalar pentadiagonal systems rather than block pentadiagonal systems translate into savings in computing time.

Results for the transonic channel flow problem are displayed in Figs. 4.15 through 4.17. The solutions again show good agreement with the exact solution, as shown in Fig. 4.15. Note in particular the manner in which the shock is captured with the

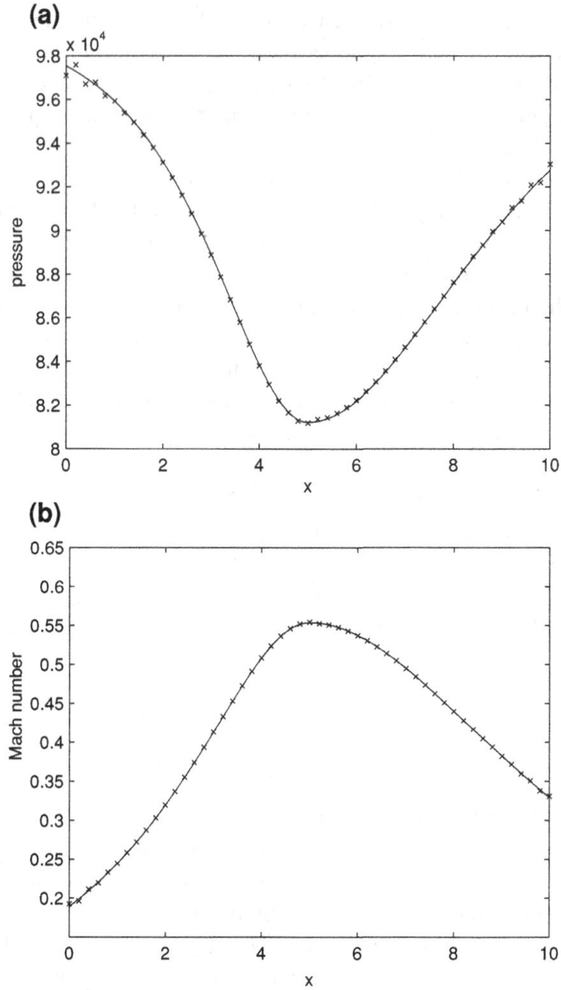

Fig. 4.9 Comparison of exact (-) solution for the subsonic channel flow problem with the numerical (x) solution computed on a grid with with 49 interior nodes

solution at one grid node lying midway between the values upstream and downstream of the shock. Figure 4.16 shows the residual convergence achieved with the block form of the algorithm at a Courant number of 120. The diagonal form proves to be unstable at a Courant number of 120 with a grid consisting of 99 interior nodes. However, at a Courant number of 70 it converges in slightly fewer iterations than the block form, as shown in Fig. 4.17.

Finally, Fig. 4.18 compares the numerical and exact solutions for the shock-tube problem on a grid with 400 cells with a maximum Courant number of unity. With the present numerical dissipation model, the shock wave and contact surface are spread out over several cells. This is the motivation for the methods described in Chap. 6.

Fig. 4.10 Comparison of exact (-) solution for the subsonic channel flow problem with the numerical (x) solution computed on a grid with 199 interior nodes

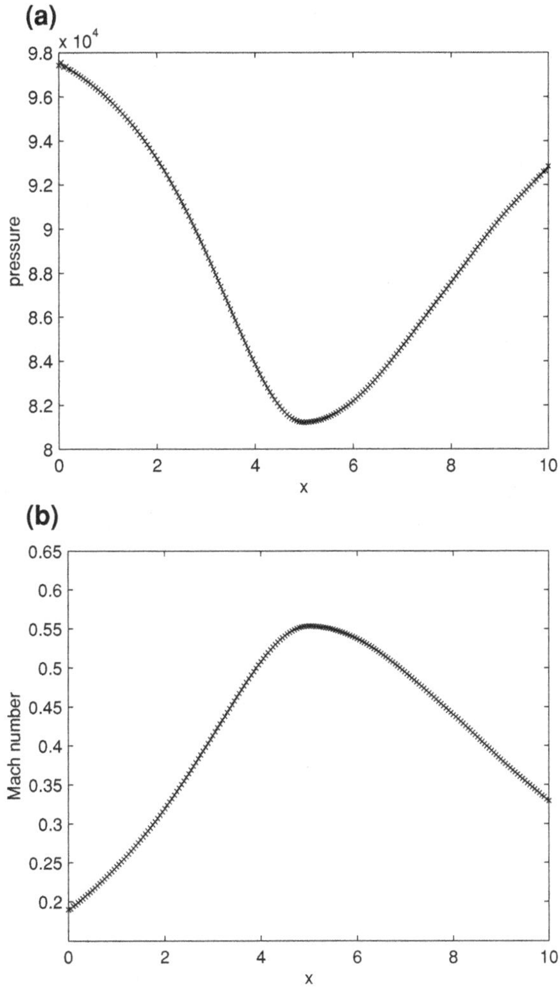

4.9 Summary

The algorithm described in this chapter has the following key features:

- The discretization of the spatial derivatives is accomplished through second-order centered difference operators applied in a uniform computational space. This is facilitated by a curvilinear coordinate transformation that is defined implicitly through a structured grid. This approach is restricted to structured or block-structured grids. Numerical dissipation is added through a nonlinear artificial dissipation scheme that combines a third-order dissipative term in smooth regions

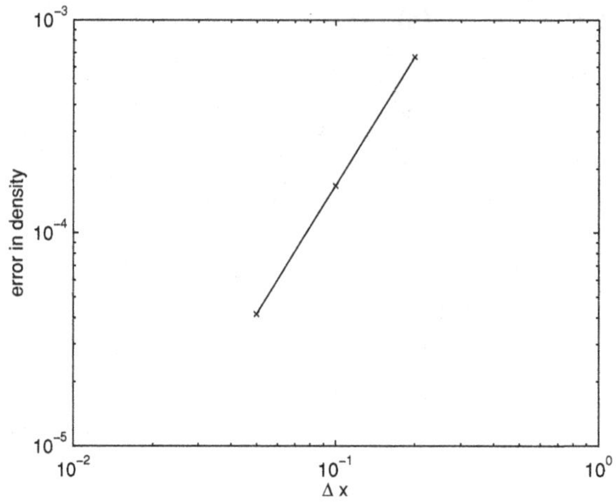

Fig. 4.11 Numerical error in density plotted versus grid spacing for the subsonic channel flow problem computed with linear extrapolation of outgoing Riemann invariants and $\kappa_2 = 0$

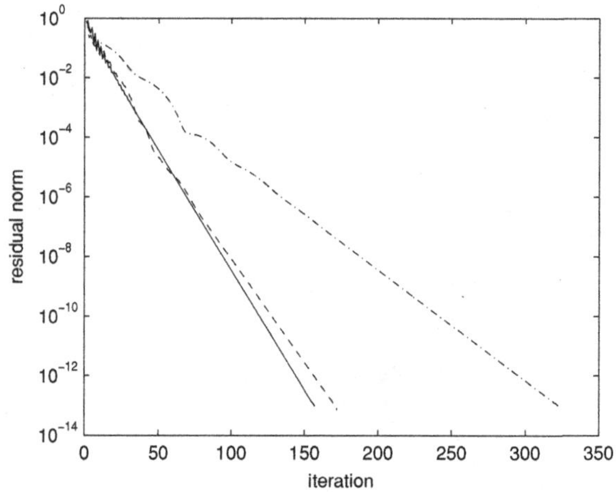

Fig. 4.12 Residual convergence histories for the subsonic channel flow problem using the block form of the implicit algorithm on a grid with 99 interior nodes with $C_n = 40$ (-), $C_n = 20$ (- -), and $C_n = 10$ (-·)

of the flow with a first-order term near shock waves. A pressure-based term is used as a shock sensor.

- After discretization in space, the original PDEs are converted to a large system of ODEs. For computations of steady flows, the implicit Euler method is used to follow a time dependent, though not time accurate, path to steady state. A local time linearization is applied, and the implicit operator is approximately factored in order to reduce the computational work required at each time step. With the

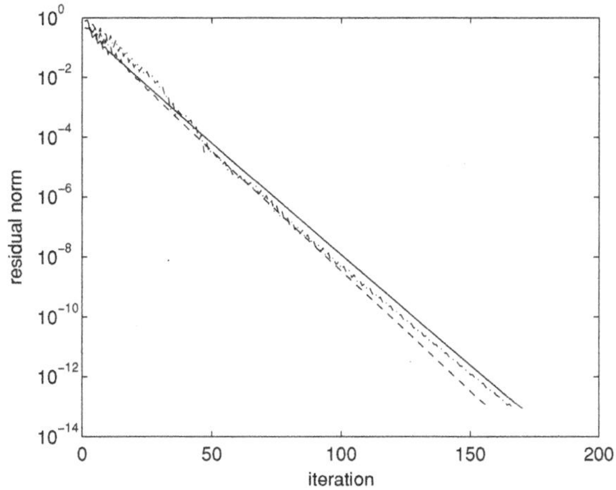

Fig. 4.13 Residual convergence histories for the subsonic channel flow problem using the block form of the implicit algorithm with $C_n = 40$ on a grid with 49 interior nodes (-), 99 interior nodes (- -), and 199 interior nodes (-·)

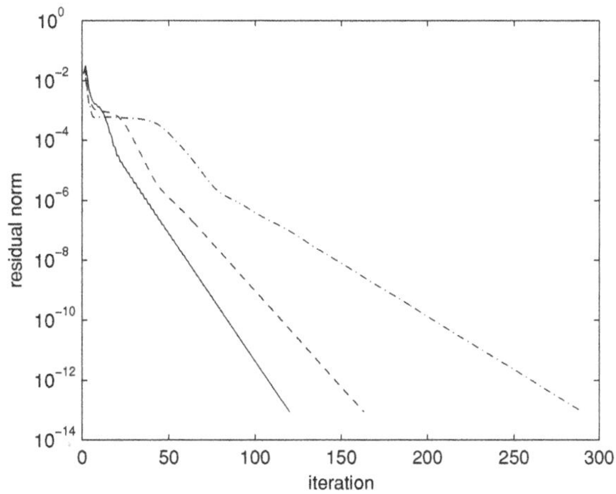

Fig. 4.14 Residual convergence histories for the subsonic channel flow problem using the diagonal form of the implicit algorithm on a grid with 99 interior nodes with $C_n = 40$ (-), $C_n = 20$ (- -), and $C_n = 10$ (-·)

approximately factored form, block pentadiagonal linear systems must be solved. The approximate factorization has a detrimental effect on the convergence rate at large time steps but greatly reduces the computational cost per time step in comparison with a direct solution technique. The cost per time step can be further reduced through the use of the diagonal form, which reduces the necessary inversions to scalar pentadiagonal matrices. Convergence can be further accelerated through local time stepping and mesh sequencing. For time-accurate computations of unsteady flows, the block form of the approximate factorization algorithm can be applied to the second-order backward or the trapezoidal implicit time-marching methods. Alternatively, the dual time stepping approach can be used where the steady form of the algorithm is used to solve the nonlinear problem arising at each implicit time step.

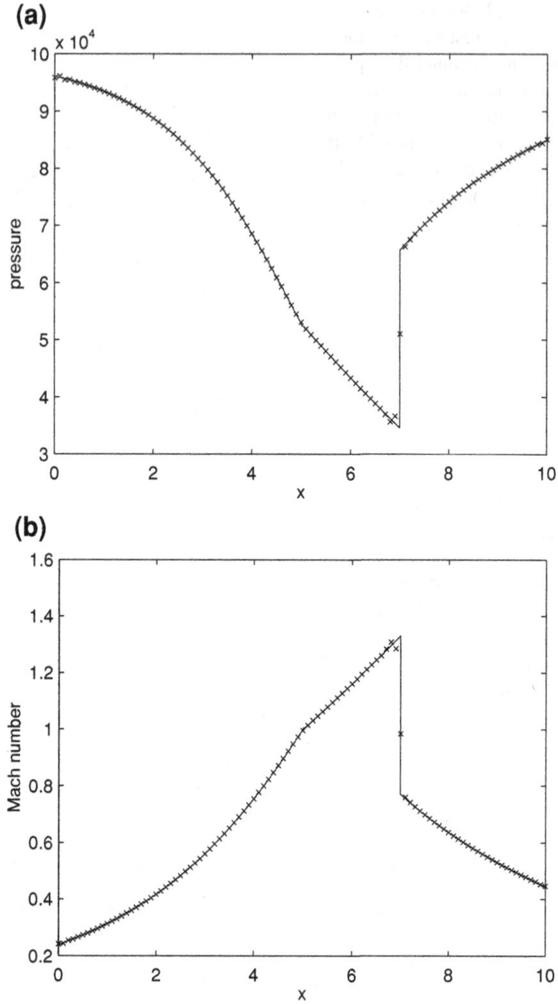

Fig. 4.15 Comparison of exact (-) solution for the transonic channel flow problem with the numerical (x) solution computed on a grid with with 99 interior nodes

4.10 Exercises

For related discussion, see Sect. 4.8.

4.1 Write a computer program to apply the implicit finite-difference algorithm presented in this chapter to the quasi-one-dimensional Euler equations for the following subsonic problem. $S(x)$ is given by

$$S(x) = \begin{cases} 1 + 1.5\left(1 - \frac{x}{5}\right)^2 & 0 \leq x \leq 5 \\ 1 + 0.5\left(1 - \frac{x}{5}\right)^2 & 5 \leq x \leq 10 \end{cases} \tag{4.177}$$

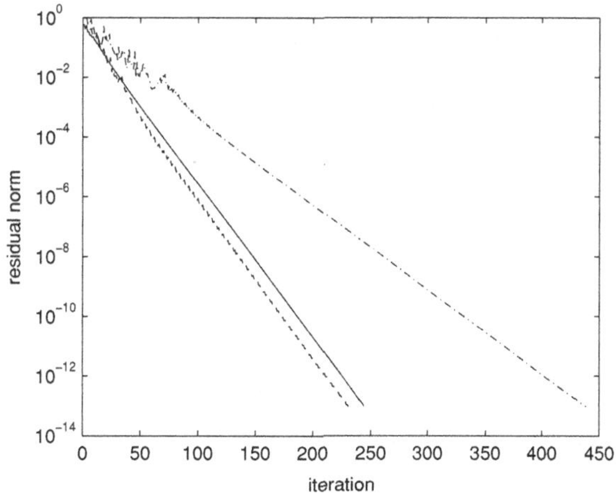

Fig. 4.16 Residual convergence histories for the transonic channel flow problem using the block form of the implicit algorithm with $C_n = 120$ on a grid with 49 interior nodes (-), 99 interior nodes (- -), and 199 interior nodes (-·)

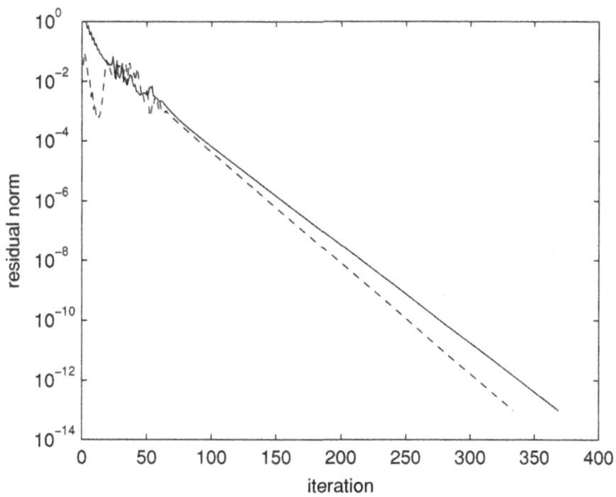

Fig. 4.17 Residual convergence histories for the transonic channel flow problem using the block form (-) and the diagonal form (- -) of the implicit algorithm with $C_n = 70$ on a grid with 99 interior nodes

where $S(x)$ and x are in meters. The fluid is air, which is considered to be a perfect gas with $R = 287$ N m kg^{-1} K^{-1}, and $\gamma = 1.4$, the total temperature is $T_0 = 300$ K, and the total pressure at the inlet is $p_{01} = 100$ kPa. The flow is subsonic throughout the channel, with $S^* = 0.8$. Use implicit Euler time marching with and without the diagonal form. Use the nonlinear scalar artificial dissipation model. Compare your solution with the exact solution computed in Exercise 3.1. Show the convergence history for each case. Experiment with parameters, such as the Courant number and the artificial dissipation coefficients, to examine their effect on convergence and accuracy.

4.2 Repeat Exercise 4.1 for a transonic flow in the same channel. The flow is subsonic at the inlet, there is a shock at $x = 7$, and $S^* = 1$. Compare your solution with that calculated in Exercise 3.2.

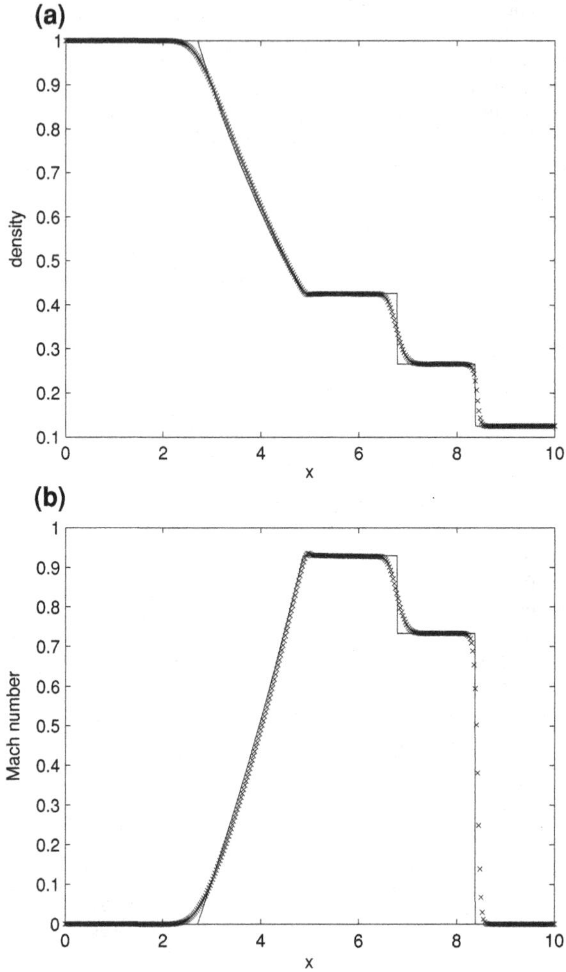

Fig. 4.18 Comparison of the exact solution (-) for the shock-tube problem at $t = 6.1$ ms with the numerical solution (x) computed on a grid with 400 cells with a maximum Courant number of unity

4.3 Write a computer program to apply the implicit finite-difference algorithm presented in this chapter to the following shock-tube problem: $p_L = 10^5$, $\rho_L = 1$, $p_R = 10^4$, and $\rho_R = 0.125$, where the pressures are in Pa and the densities in kg/m^3. The fluid is a perfect gas with $\gamma = 1.4$. Use both implicit Euler and second-order backwards time marching with and without the diagonal form. Compare your solution at $t = 6.1$ ms with that found in Exercise 3.3. Examine the effect of the time step and the artificial dissipation parameters on the accuracy of the solution.

Appendix: Flux Jacobian Eigensystems in Two and Three Dimensions

The flux Jacobian matrices of Eq. 4.104 have real eigenvalues and a complete set of eigenvectors. The similarity transforms are

$$\widehat{A} = T_\xi \Lambda_\xi T_\xi^{-1} \quad \text{and} \quad \widehat{B} = T_\eta \Lambda_\eta T_\eta^{-1}. \tag{4.178}$$

where

$$\Lambda_\xi = \begin{bmatrix} U & & & \\ & U & & \\ & & U + a\sqrt{\xi_x^2 + \xi_y^2} & \\ & & & U - a\sqrt{\xi_x^2 + \xi_y^2} \end{bmatrix} \tag{4.179}$$

$$\Lambda_\eta = \begin{bmatrix} V & & & \\ & V & & \\ & & V + a\sqrt{\eta_x^2 + \eta_y^2} & \\ & & & V - a\sqrt{\eta_x^2 + \eta_y^2} \end{bmatrix}, \tag{4.180}$$

with

$$T_\kappa = \begin{bmatrix} 1 & 0 & \alpha & \alpha \\ u & \widetilde{\kappa}_y \rho & \alpha(u + \widetilde{\kappa}_x a) & \alpha(u - \widetilde{\kappa}_x a) \\ v & -\widetilde{\kappa}_x \rho & \alpha(v + \widetilde{\kappa}_y a) & \alpha(v - \widetilde{\kappa}_y a) \\ \frac{\phi^2}{(\gamma - 1)} & \rho(\widetilde{\kappa}_y u - \widetilde{\kappa}_x v) & \alpha\left[\frac{\phi^2 + a^2}{(\gamma - 1)} + a\widetilde{\theta}\right] & \alpha\left[\frac{\phi^2 + a^2}{(\gamma - 1)} - a\widetilde{\theta}\right] \end{bmatrix} \tag{4.181}$$

$$T_\kappa^{-1} = \begin{bmatrix} (1 - \phi^2/a^2) & (\gamma - 1)u/a^2 & (\gamma - 1)v/a^2 & -(\gamma - 1)/a^2 \\ -(\widetilde{\kappa}_y u - \widetilde{\kappa}_x v)/\rho & \widetilde{\kappa}_y/\rho & -\widetilde{\kappa}_x/\rho & 0 \\ \beta(\phi^2 - a\widetilde{\theta}) & \beta[\widetilde{\kappa}_x a - (\gamma - 1)u] & \beta[\widetilde{\kappa}_y a - (\gamma - 1)v] & \beta(\gamma - 1) \\ \beta(\phi^2 + a\widetilde{\theta}) & -\beta[\widetilde{\kappa}_x a + (\gamma - 1)u] & -\beta[\widetilde{\kappa}_y a + (\gamma - 1)v] & \beta(\gamma - 1) \end{bmatrix}, \tag{4.182}$$

and $\alpha = \rho/(\sqrt{2}a)$, $\beta = 1/(\sqrt{2}\rho a)$, $\widetilde{\theta} = \widetilde{\kappa}_x u + \widetilde{\kappa}_y v, \phi = \frac{1}{2}(\gamma - 1)(u^2 + v^2)$ and, for example, $\widetilde{\kappa}_x = \kappa_x/\sqrt{\kappa_x^2 + \kappa_y^2}$.

Relations exist between T_ξ and T_η of the form

$$\widehat{N} = T_\xi^{-1} T_\eta, \quad \widehat{N}^{-1} = T_\eta^{-1} T_\xi, \tag{4.183}$$

where

$$\widehat{N} = \begin{bmatrix} 1 & 0 & 0 & 0 \\ 0 & m_1 & -\mu m_2 & \mu m_2 \\ 0 & \mu m_2 & \mu^2(1+m_1) & \mu^2(1-m_1) \\ 0 & -\mu m_2 & \mu^2(1-m_1) & \mu^2(1+m_1) \end{bmatrix}, \tag{4.184}$$

and

$$\widehat{N}^{-1} = \begin{bmatrix} 1 & 0 & 0 & 0 \\ 0 & m_1 & \mu m_2 & -\mu m_2 \\ 0 & -\mu m_2 & \mu^2(1+m_1) & \mu^2(1-m_1) \\ 0 & \mu m_2 & \mu^2(1-m_1) & \mu^2(1+m_1) \end{bmatrix}, \tag{4.185}$$

with $m_1 = \left(\widetilde{\xi}_x \, \widetilde{\eta}_x + \widetilde{\xi}_y \widetilde{\eta}_y\right)$, $m_2 = \left(\widetilde{\xi}_x \, \widetilde{\eta}_y - \widetilde{\xi}_y \widetilde{\eta}_x\right)$ and $\mu = 1/\sqrt{2}$. It is interesting to note that the matrix $\widehat{N}$ is only a function of the metrics and not the flow variables.

In three dimensions the Jacobian matrices $\widehat{A}$, $\widehat{B}$, or $\widehat{C}$ =

$$\left[\begin{array}{cc} \kappa_t & \kappa_x \\ \kappa_x \phi^2 - u\theta & \kappa_t + \theta - \kappa_x(\gamma-2)u \\ \kappa_y \phi^2 - v\theta & \kappa_x v - \kappa_y(\gamma-1)u \\ \kappa_z \phi^2 - w\theta & \kappa_x w - \kappa_z(\gamma-1)u \\ -\theta\left(\gamma e/\rho - 2\phi^2\right) & \kappa_x\left(\gamma e/\rho - \phi^2\right) - (\gamma-1)u\theta \end{array}\right.$$

$$\left.\begin{array}{ccc} \kappa_y & \kappa_z & 0 \\ \kappa_y u - \kappa_x(\gamma-1)v & \kappa_z u - \kappa_x(\gamma-1)w & \kappa_x(\gamma-1) \\ \kappa_t + \theta - \kappa_y(\gamma-2)v & \kappa_z v - \kappa_y(\gamma-1)w & \kappa_y(\gamma-1) \\ \kappa_y w - \kappa_z(\gamma-1)v & \kappa_t + \theta - \kappa_z(\gamma-2)w & \kappa_z(\gamma-1) \\ \kappa_y\left(\gamma e\rho^{-1} - \phi^2\right) - (\gamma-1)v\theta & \kappa_z\left(\gamma e\rho^{-1} - \phi^2\right) - (\gamma-1)w\theta & \kappa_t + \gamma\theta \end{array}\right], \tag{4.186}$$

where

$$\begin{aligned} \theta &= \kappa_x u + \kappa_y v + \kappa_z w \\ \phi^2 &= (\gamma-1)\left(\frac{u^2+v^2+w^2}{2}\right), \end{aligned} \tag{4.187}$$

with $\kappa = \xi$, η, or ζ for $\widehat{A}$, $\widehat{B}$, or $\widehat{C}$, respectively.

The viscous flux Jacobian is

$$\widehat{M} = J^{-1} \begin{bmatrix} 0 & 0 & 0 & 0 & 0 \\ m_{21} & \alpha_1 \partial_\zeta(\rho^{-1}) & \alpha_2 \partial_\zeta(\rho^{-1}) & \alpha_3 \partial_\zeta(\rho^{-1}) & 0 \\ m_{31} & \alpha_2 \partial_\zeta(\rho^{-1}) & \alpha_4 \partial_\zeta(\rho^{-1}) & \alpha_5 \partial_\zeta(\rho^{-1}) & 0 \\ m_{41} & \alpha_3 \partial_\zeta(\rho^{-1}) & \alpha_5 \partial_\zeta(\rho^{-1}) & \alpha_6 \partial_\zeta(\rho^{-1}) & 0 \\ m_{51} & m_{52} & m_{53} & m_{54} & \alpha_0 \partial_\zeta(\rho^{-1}) \end{bmatrix} J, \quad (4.188)$$

where

$$\begin{aligned}
m_{21} &= -\alpha_1 \partial_\zeta(u/\rho) - \alpha_2 \partial_\zeta(v/\rho) - \alpha_3 \partial_\zeta(w/\rho) \\
m_{31} &= -\alpha_2 \partial_\zeta(u/\rho) - \alpha_4 \partial_\zeta(v/\rho) - \alpha_5 \partial_\zeta(w/\rho) \\
m_{41} &= -\alpha_3 \partial_\zeta(u/\rho) - \alpha_5 \partial_\zeta(v/\rho) - \alpha_6 \partial_\zeta(w/\rho) \\
m_{51} &= \alpha_0 \partial_\zeta \left[-(e/\rho^2) + (u^2 + v^2 + w^2)/\rho \right] \\
&\quad -\alpha_1 \partial_\zeta(u^2/\rho) - \alpha_4 \partial_\zeta(v^2/\rho) - \alpha_6 \partial_\zeta(w^2/\rho) \\
&\quad -2\alpha_2 \partial_\zeta(uv/\rho) - 2\alpha_3 \partial_\zeta(uw/\rho) - 2\alpha_5 \partial_\zeta(vw/\rho) \\
m_{52} &= -\alpha_0 \partial_\zeta(u/\rho) - m_{21}, \qquad m_{53} = -\alpha_0 \partial_\zeta(v/\rho) - m_{31} \\
m_{54} &= -\alpha_0 \partial_\zeta(w/\rho) - m_{41}, \qquad m_{44} = \alpha_4 \partial_\zeta(\rho^{-1}) \\
\alpha_0 &= \gamma \mu Pr^{-1}({\zeta_x}^2 + {\zeta_y}^2 + {\zeta_z}^2), \quad \alpha_1 = \mu[(4/3){\zeta_x}^2 + {\zeta_y}^2 + {\zeta_z}^2] \\
\alpha_2 &= (\mu/3)\zeta_x \zeta_y, \quad \alpha_3 = (\mu/3)\zeta_x \zeta_z, \quad \alpha_4 = \mu[{\zeta_x}^2 + (4/3){\zeta_y}^2 + {\zeta_z}^2] \\
\alpha_5 &= (\mu/3)\zeta_y \zeta_z, \qquad \alpha_6 = \mu[{\zeta_x}^2 + {\zeta_y}^2 + (4/3){\zeta_z}^2]. \qquad (4.189)
\end{aligned}$$

The eigensystem decompositions of the three-dimensional Jacobians have the form $\widehat{A} = T_\xi \Lambda_\xi T_\xi^{-1}$, $\widehat{B} = T_\eta \Lambda_\eta T_\eta^{-1}$, and $\widehat{C} = T_\zeta \Lambda_\zeta T_\zeta^{-1}$. The eigenvalues are

$$\begin{aligned}
\lambda_1 &= \lambda_2 = \lambda_3 = \kappa_t + \kappa_x u + \kappa_y v + \kappa_z w \\
\lambda_4 &= \lambda_1 + \kappa a, \quad \lambda_5 = \lambda_1 - \kappa a \\
\kappa &= \sqrt{\kappa_x^2 + \kappa_y^2 + \kappa_z^2}. \qquad (4.190)
\end{aligned}$$

The matrix T_κ, representing the left eigenvectors, is

$$T_\kappa = \begin{bmatrix} \tilde{\kappa}_x & \tilde{\kappa}_y & \tilde{\kappa}_z & \alpha & \alpha \\ \tilde{\kappa}_x u & \tilde{\kappa}_y u - \tilde{\kappa}_z \rho & \tilde{\kappa}_z u + \tilde{\kappa}_y \rho & \alpha(u + \tilde{\kappa}_x a) & \alpha(u - \tilde{\kappa}_x a) \\ \tilde{\kappa}_x v + \tilde{\kappa}_z \rho & \tilde{\kappa}_y u & \tilde{\kappa}_z v - \tilde{\kappa}_x \rho & \alpha(v + \tilde{\kappa}_y a) & \alpha(v - \tilde{\kappa}_y a) \\ \tilde{\kappa}_x w + \tilde{\kappa}_y \rho & \tilde{\kappa}_y w + \tilde{\kappa}_x \rho & \tilde{\kappa}_z w & \alpha(w + \tilde{\kappa}_z a) & \alpha(w - \tilde{\kappa}_z a) \\ \left[\tilde{\kappa}_x \phi^2/(\gamma-1) + \rho(\tilde{\kappa}_z v - \tilde{\kappa}_y w)\right] & \left[\tilde{\kappa}_y \phi^2/(\gamma-1) + \rho(\tilde{\kappa}_x w - \tilde{\kappa}_z u)\right] & \left[\tilde{\kappa}_z \phi^2/(\gamma-1) + \rho(\tilde{\kappa}_y u - \tilde{\kappa}_x v)\right] & \alpha\left[(\phi^2 + a^2)/(\gamma-1) + \tilde{\theta} a\right] & \alpha\left[(\phi^2 + a^2)/(\gamma-1) + \tilde{\theta} a\right] \end{bmatrix}, \tag{4.191}$$

where

$$\alpha = \frac{\rho}{\sqrt{2}a}, \quad \tilde{\kappa}_x = \frac{\kappa_x}{\kappa}, \quad \tilde{\kappa}_y = \frac{\kappa_y}{\kappa}, \quad \tilde{\kappa}_z = \frac{\kappa_z}{\kappa}, \quad \tilde{\theta} = \frac{\theta}{\kappa}. \tag{4.192}$$

The corresponding T_κ^{-1} is

$$T_\kappa^{-1} = \begin{bmatrix} \tilde{\kappa}_x(1 - \phi^2/a^2) - (\tilde{\kappa}_z v - \tilde{\kappa}_y w)/\rho & \tilde{\kappa}_x(\gamma-1)u/a^2 & \tilde{\kappa}_x(\gamma-1)v/a^2 + \tilde{\kappa}_z/\rho & \tilde{\kappa}_x(\gamma-1)w/a^2 - \tilde{\kappa}_y/\rho & -\tilde{\kappa}_x(\gamma-1)/a^2 \\ \tilde{\kappa}_y(1 - \phi^2/a^2) - (\tilde{\kappa}_x w - \tilde{\kappa}_z u)/\rho & \tilde{\kappa}_y(\gamma-1)u/a^2 - \tilde{\kappa}_z/\rho & \tilde{\kappa}_y(\gamma-1)v/a^2 & \tilde{\kappa}_y(\gamma-1)w/a^2 + \tilde{\kappa}_x/\rho & -\tilde{\kappa}_y(\gamma-1)/a^2 \\ \tilde{\kappa}_z(1 - \phi^2/a^2) - (\tilde{\kappa}_y u - \tilde{\kappa}_x v)/\rho & \tilde{\kappa}_z(\gamma-1)u/a^2 + \tilde{\kappa}_y/\rho & \tilde{\kappa}_z(\gamma-1)v/a^2 - \tilde{\kappa}_x/\rho & \tilde{\kappa}_z(\gamma-1)w/a^2 & -\tilde{\kappa}_z(\gamma-1)/a^2 \\ \beta(\phi^2 - \tilde{\theta} a) & -\beta[(\gamma-1)u - \tilde{\kappa}_x a] & -\beta[(\gamma-1)v - \tilde{\kappa}_y a] & -\beta[(\gamma-1)w - \tilde{\kappa}_z a] & \beta(\gamma-1) \\ \beta(\phi^2 + \tilde{\theta} a) & -\beta[(\gamma-1)u + \tilde{\kappa}_x a] & -\beta[(\gamma-1)v + \tilde{\kappa}_y a] & -\beta[(\gamma-1)w + \tilde{\kappa}_z a] & \beta(\gamma-1) \end{bmatrix}, \tag{4.193}$$

where

$$\beta = \frac{1}{\sqrt{2}\rho a}. \tag{4.194}$$

References

1. Beam, R.M., Warming, R.F.: An implicit finite-difference algorithm for hyperbolic systems in conservation law form. J. Comput. Phys. **22**, 87–110 (1976)
2. Steger, J.L.: Implicit finite difference simulation of flow about arbitrary geometries with application to airfoils. AIAA Paper 77-665 (1977)
3. Warming, R.F., Beam, R.M.: On the construction and application of implicit factored schemes for conservation laws. In: SIAM-AMS Proceedings, vol. 11 (1978)
4. Pulliam, T.H., Steger, J.L.: Implicit finite-difference simulations of three dimensional compressible flow. AIAA J. **18**, 159–167 (1980)
5. Pulliam, T.H., Chaussee, D.S.: A diagonal form of an implicit approximate factorization algorithm. J. Comput. Phys. **39**, 347–363 (1981)
6. Pulliam, T.H.: Efficient solution methods for the Navier-Stokes equations. In: Von Karman Institute for Fluid Dynamics Numerical Techniques for Viscous Flow Calculations in Turbomachinery Bladings (1986)
7. Viviand, H.: Formes conservatives des équations de la dynamique des gaz. Recherche Aérospatiale **1**, 65–66 (1974)
8. Vinokur, M.: Conservation equations of gasdynamics in curvilinear coordinate systems. J. Comput. Phys. **14**, 105–125 (1974)
9. Baldwin, B.S., Lomax, H.: Thin-layer approximation and algebraic model for separated turbulent flows. AIAA Paper 78-257 (1978)
10. Gustafsson, B.: The convergence rate for difference approximations to mixed initial boundary value problems. Math. Comput. **29**, 396–406 (1975)
11. Thomas, P.D., Lombard, C.K.: Geometric conservation law and its application to flow computations on moving grids. AIAA J. **17**, 1030–1037 (1979)
12. Jameson, A., Schmidt, W., Turkel, E.: Numerical solutions of the Euler equations by finite volume methods using Runge-Kutta time-stepping schemes. AIAA Paper 81-1259 (1981)
13. Lomax, H., Pulliam, T.H., Zingg, D.W.: Fundamentals of Computational Fluid Dynamics. Springer, Berlin (2001)
14. Pulliam, T.H.: Artificial dissipation models for the Euler equations. AIAA J. **24**, 1931–1940 (1986)
15. Saad, Y., Schultz, M.H.: A generalized minimal residual algorithm for solving nonsymmetric linear systems. SIAM J. Sci. Stat. Comput. **7**, 856–869 (1986)
16. Warming, R.F., Beam, R.M., Hyett, B.J.: Diagonalization and simultaneous symmetrization of the gas-dynamic matrices. Math. Comput. **29**, 1037–1045 (1975)
17. Chakravarty, S.: Euler equations-implicit schemes and implicit boundary conditions. AIAA Paper 82-0228 (1982)
18. Colonius, T., Lele, S.K.: Computational aeroacoustics: progress on nonlinear problems of sound generation. Prog. Aerosp. Sci. **40**, 345–416 (2004)
19. Svard, M., Carpenter, M.H., Nordström, J.: A stable high-order finite difference scheme for the compressible Navier-Stokes equations, far-field boundary conditions. J. Comput. Phys. **225**, 1020–1038 (2007)
20. Osusky, M., Zingg, D.W.: Parallel Newton-Krylov-Schur Flow Solver for the Navier-Stokes Equations. AIAA J. **51**, 2833–2851 (2013)
21. Salas, M., Jameson, A., Melnik, R.A.: Comparative study of the nonuniqueness problem of the potential equation. AIAA Paper 83-1888 (1983)

Chapter 5
An Explicit Finite-Volume Algorithm with Multigrid

5.1 Introduction

The salient features of the algorithm presented in this chapter are as follows (the reader is urged to contrast these with the key characteristics of the algorithm presented in Chap. 4, which are listed in Sect. 4.1):

- cell-centered data storage; the numerical solution for the state variables is associated with the cells of the grid
- second-order finite-volume spatial discretization with added numerical dissipation; a simple shock-capturing device
- applicable to structured grids (see Sect. 4.2)
- explicit multi-stage time marching with implicit residual smoothing and multigrid

Key contributions to the development of this algorithm were made by Jameson et al. [1], Baker et al. [2], Jameson and Baker [3], Jameson [4, 5], and Swanson and Turkel [6, 7]. The reader is referred to Swanson and Turkel [7] for further analysis and description of the algorithm.

The exercises at the end of this chapter again provide an opportunity to apply the algorithm presented to several one-dimensional problems.

5.2 Spatial Discretization: Cell-Centered Finite-Volume Method

The cell-centered approach contrasts with the node-centered approach described in Chap. 4. The meshes described thus far are known as *primary* meshes. One can also construct a *dual* mesh by joining the centroids of the cells associated with the primary mesh. In the case of a two-dimensional structured mesh, the dual mesh also consists of quadrilaterals and is qualitatively similar to the primary mesh. For more general unstructured meshes this is not the case. For example, for a primary mesh consisting of regular triangles the dual mesh consists of hexagons. A scheme that is

T. H. Pulliam and D. W. Zingg, *Fundamental Algorithms in Computational Fluid Dynamics*, Scientific Computation, DOI: 10.1007/978-3-319-05053-9_5,

cell centered on the primary mesh can be considered to be node centered on the dual mesh. Hence, in the case of quadrilateral structured meshes, the cell-centered nature has little impact on the spatial discretization in the interior, and both cell-centered and node-centered finite-volume schemes are in common use on both structured and unstructured meshes. The main differences between the two arise at boundaries and in the construction of coarse meshes for multigrid. This will be discussed further below.

A finite-volume method numerically solves the governing equations in integral form, as presented in Sect. 3.1.2. In their most general coordinate-free form, conservation laws can be written as

$$\frac{\mathrm{d}}{\mathrm{d}t}\int_{V(t)} Q\mathrm{d}V + \oint_{S(t)} \hat{n}\cdot\mathcal{F}\mathrm{d}S = \int_{V(t)} P\mathrm{d}V, \tag{5.1}$$

where P is a source term, and the other variables are defined in Chap. 3. If we restrict our interest to two-dimensional problems without source terms and meshes that are static with respect to time, we obtain

$$\frac{\mathrm{d}}{\mathrm{d}t}\int_{A} Q\mathrm{d}A + \oint_{C} \hat{n}\cdot\mathcal{F}\mathrm{d}l = 0, \tag{5.2}$$

where A is a control volume bounded by a contour C. Writing the flux tensor $\mathcal{F}$ in Cartesian coordinates and separating inviscid and viscous fluxes gives

$$\frac{\mathrm{d}}{\mathrm{d}t}\int_{A} Q\mathrm{d}A + \oint_{C} \hat{n}\cdot(E\hat{i}+F\hat{j})\mathrm{d}l = \oint_{C} \hat{n}\cdot(E_{\mathrm{v}}\hat{i}+F_{\mathrm{v}}\hat{j})\mathrm{d}l. \tag{5.3}$$

Finally, writing the product of the outward normal and the length of the cell edge in Cartesian coordinates as

$$\hat{n}\mathrm{d}l = \mathrm{d}y\hat{i} - \mathrm{d}x\hat{j} \tag{5.4}$$

gives the final form to be discretized using the finite volume method:

$$\frac{\mathrm{d}}{\mathrm{d}t}\int_{A} Q\mathrm{d}A + \oint_{C}(E\mathrm{d}y - F\mathrm{d}x) = \oint_{C}(E_{\mathrm{v}}\mathrm{d}y - F_{\mathrm{v}}\mathrm{d}x). \tag{5.5}$$

The semi-discrete form of (5.5) is written as

$$A_{j,k}\frac{\mathrm{d}}{\mathrm{d}t}Q_{j,k} + \mathcal{L}_{\mathrm{i}}Q_{j,k} + \mathcal{L}_{\mathrm{ad}}Q_{j,k} = \mathcal{L}_{\mathrm{v}}Q_{j,k}, \tag{5.6}$$

where $A_{j,k}$ is the area of the cell, $\mathcal{L}_{\mathrm{i}}$ is the discrete approximation to the inviscid flux integral, $\mathcal{L}_{\mathrm{ad}}$ is the artificial dissipation operator, $\mathcal{L}_{\mathrm{v}}$ is the discrete approximation to the viscous flux integral, and $Q_{j,k}$ denotes the conservative variables averaged over cell j, k as follows:

$$Q_{j,k} = \frac{1}{A_{j,k}} \int_{A_{j,k}} Q \mathrm{d}A. \tag{5.7}$$

The terms $\mathcal{L}_\mathrm{i} Q$, $\mathcal{L}_\mathrm{v} Q$, and $\mathcal{L}_\mathrm{ad} Q$ are described next.

5.2.1 Inviscid and Viscous Fluxes

The inviscid flux integral is approximated by summing over the four edges of the cell as follows:

$$\mathcal{L}_\mathrm{i} Q = \sum_{l=1}^{4} (\mathcal{F}_\mathrm{i})_l \cdot \mathbf{s}_l, \tag{5.8}$$

where

$$\mathbf{s}_l = (\Delta y)_l \hat{i} - (\Delta x)_l \hat{j}. \tag{5.9}$$

is the discrete analog of (5.4) for straight cell edges, and $(\mathcal{F}_\mathrm{i})_l$ is an approximation to the inviscid flux tensor at the cell edge. We use boldface to emphasize that $\mathbf{s}_l$ is a vector. The terms $(\Delta x)_l$ and $(\Delta y)_l$ must be defined such that the normal vector points out of the cell. Since the cell edges are straight, the outward normal is constant along each edge. The only exception might arise at the body surface; there the approximation of the edge as straight is adequate for a second-order discretization, but the curvature of the boundary must be taken into account if higher-order accuracy is desired.

In Sect. 2.4.2, we saw that the combination of a piecewise constant reconstruction with a simple average for resolving the discontinuity in fluxes at cell interfaces leads to a second-order centered finite-volume scheme that is analogous to a second-order centered finite-difference scheme on a uniform mesh. The same approach is taken here. With the minus sign superscript defining quantities in the cell on one side of the interface and the plus sign indicating the other side, the averaged flux on a given cell edge is given by

$$(\mathcal{F}_\mathrm{i})_l = \frac{1}{2}(\mathcal{F}_\mathrm{i}^- + \mathcal{F}_\mathrm{i}^+) = \frac{1}{2}(Q^- \mathbf{v}^- + Q^+ \mathbf{v}^+)_l + \bar{\mathcal{P}}_l, \tag{5.10}$$

where $\mathbf{v} = u\hat{i} + v\hat{j}$, and

$$\bar{\mathcal{P}}_l = [\, 0,\ \ \frac{1}{2}(p^- + p^+)_l \hat{i},\ \ \frac{1}{2}(p^- + p^+)_l \hat{j},\ \ \frac{1}{2}(p^- \mathbf{v}^- + p^+ \mathbf{v}^+)_l \,]^T. \tag{5.11}$$

This scheme is second order and nondissipative. Numerical dissipation must be added, as described in Sect. 5.2.2.

For the viscous terms, we turn again to Sect. 2.4.2. In that section, a second-order finite-volume scheme was derived for the diffusion equation using two different approaches. The first approach is based on the use of a one-dimensional version of (2.58), which is given by

$$\int_A \nabla Q \mathrm{d}A = \oint_C \hat{n} Q \mathrm{d}l. \tag{5.12}$$

This approach is simple to extend to multidimensions but is restricted to second-order accuracy. Given that we seek a second-order approximation, we will follow this approach to obtain a discretization for the viscous flux terms.

The difficulty associated with the viscous fluxes is that they include velocity gradients, and these cannot be obtained directly from the solution vector. In order to obtain a suitable approximation to the velocity gradients at the cell edges, (5.12) is applied to auxiliary cells that surround each edge of the cell in question. When applied to the Cartesian velocity components, (5.12) gives the components of the velocity gradient as follows:

$$\begin{aligned} \int_{A'} \frac{\partial u}{\partial x} \mathrm{d}A &= \oint_{C'} u \mathrm{d}y \\ \int_{A'} \frac{\partial u}{\partial y} \mathrm{d}A &= -\oint_{C'} u \mathrm{d}x, \end{aligned} \tag{5.13}$$

with analogous expressions for the components of the gradient of v, where the primes are added to remind the reader that these expressions are used for auxiliary cells surrounding the edges of the finite volume. A second-order approximation to the integrals on the right-hand side of these expressions divided by the cell area provides an approximation to the average gradient in the cell. This then provides an approximation, valid to second order, to the gradient along the edge contained in the auxiliary cell.

A sample auxiliary cell is depicted in Fig. 5.1 [7]. The cell in question is cell j, k defined by ABCD. The auxiliary cell A′B′C′D′ provides the approximation to the velocity gradient on edge BC. In order to evaluate the integrals on the right-hand side of (5.13), the midpoint rule is applied on each edge of cell A′B′C′D′. The velocity at the midpoint of edge A′B′ is taken as the average of the velocities associated with the four cells surrounding this edge. The same applies to edge C′D′. The velocity at the midpoint of edge B′C′ is simply that associated with cell $j, k+1$, while the velocity on edge D′A′ is that associated with cell j, k. Once the velocity gradients are approximated, all other quantities needed to form the viscous fluxes on the edges of cell j, k, including the viscosity, are obtained by averaging the quantities associated with the cells on either side of the edge in question.

An alternative auxiliary cell can be formed with the vertices being the end points of the edge and the centroids of the cells on either side of the edge, sometimes called

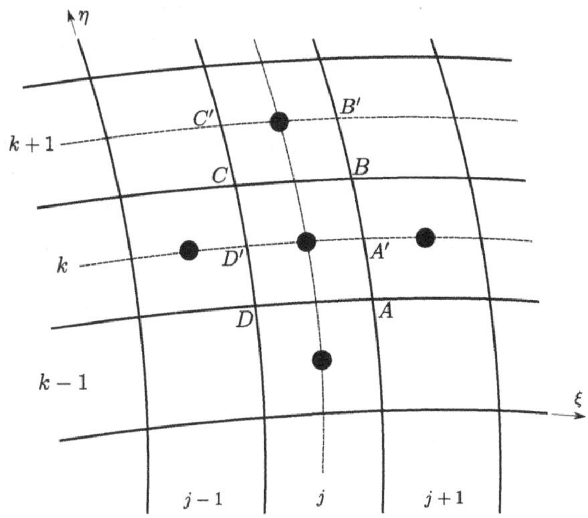

Fig. 5.1 Auxiliary cell A′B′C′D′ for computing viscous fluxes

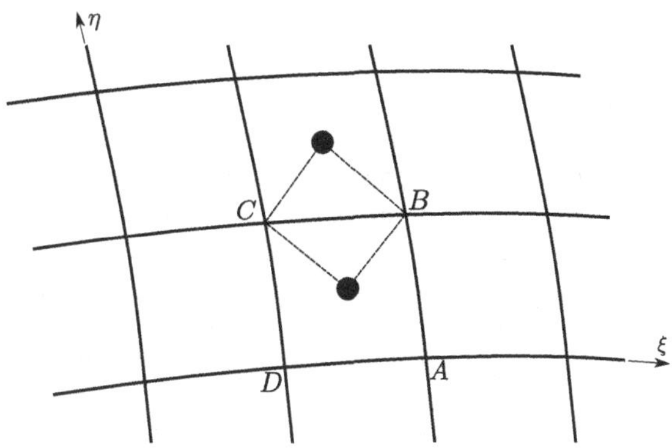

Fig. 5.2 Alternative auxiliary cell based on diamond path

a diamond path, as shown in Fig. 5.2. In this case, the trapezoidal rule is used for the integration to calculate the velocity gradients.

Once the viscous flux tensor $(\mathcal{F}_{\mathrm{v}})_l$ has been approximated at the cell edges, the net flux is determined from

$$\mathcal{L}_{\mathrm{v}} Q = \sum_{l=1}^{4} (\mathcal{F}_{\mathrm{v}})_l \cdot \mathbf{s}_l. \tag{5.14}$$

5.2.2 *Artificial Dissipation*

In analogy to the inviscid fluxes, we write the dissipation model in the following form:

$$\mathcal{L}_{\text{ad}} Q = \sum_{l=1}^{4} \mathcal{D}_l \cdot \mathbf{s}_l, \tag{5.15}$$

where $\mathcal{D}_l$ is the numerical dissipation tensor associated with each cell edge. We exploit the fact that the algorithm is applied to structured meshes. Although there is no coordinate transformation, there are effectively two coordinate directions ξ and η associated with each cell, as depicted in Fig. 5.1. Hence there are two opposing cell edges along which η varies but ξ does not, and there are two opposing cell edges along which ξ varies but η does not. For the two edges at constant ξ, the artificial dissipation tensor is given by

$$\mathcal{D} = -\epsilon^{(2)}(|A\hat{i} + B\hat{j}|)\Delta_\xi Q + \epsilon^{(4)}(|A\hat{i} + B\hat{j}|)\Delta_\xi \nabla_\xi \Delta_\xi Q, \tag{5.16}$$

where the superscripts (2) and (4) denote second- and fourth-difference dissipation, respectively, the meaning of $|A\hat{i} + B\hat{j}|$ is consistent with (2.103), A and B are the Jacobians of the inviscid flux vectors E and F, and Δ_ξ and ∇_ξ represent undivided differences in the ξ direction. For example, $\Delta_\xi Q$ is the difference between the Q values in the cells on either side of the edge. The coefficients $\epsilon^{(2)}$ and $\epsilon^{(4)}$ control the relative contribution from the two terms, analogous to the artificial dissipation scheme described in Chap. 4, and are defined below.

The reader should observe the similarity between (5.16) and (4.85). The artificial dissipation scheme described in this section is a finite-volume analog to the scheme presented in Sect. 4.4.3. Therefore it has the same basic properties. For example, the second-difference term is first order and is used near shocks, while the fourth-difference term is third order and is used in smooth regions of the flow.

Substituting the definition of $\mathbf{s}_l$ given in (5.9), we obtain for the two edges with constant ξ

$$\mathcal{D}_l \cdot \mathbf{s}_l = -\epsilon_l^{(2)}(|A_l \Delta y_l - B_l \Delta x_l|)\Delta_\xi Q + \epsilon_l^{(4)}(|A_l \Delta y_l - B_l \Delta x_l|)\Delta_\xi \nabla_\xi \Delta_\xi Q. \tag{5.17}$$

The flux Jacobians are based on an average of the two states on either side of the edge. The Roe average (Sect. 6.3) can be used. A scalar form is obtained as follows:

$$\mathcal{D}_l \cdot \mathbf{s}_l = -\epsilon_l^{(2)}(\lambda_\xi)_l \Delta_\xi Q + \epsilon_l^{(4)}(\lambda_\xi)_l \Delta_\xi \nabla_\xi \Delta_\xi Q, \tag{5.18}$$

where

$$\lambda_\xi = |u\Delta y - v\Delta x| + a\sqrt{\Delta y^2 + \Delta x^2} \tag{5.19}$$

is the appropriate spectral radius for edges of constant ξ (see Warming et al. [8]). The spectral radius term in the η direction has the same form, but the values of Δx and Δy are associated with edges of constant η.

The treatment of the pressure sensor is consistent with (4.83), giving for the edge $j + \frac{1}{2}, k$:

$$\begin{aligned}
\epsilon_l^{(2)} &= \kappa_2 \max(\Upsilon_{j+2,k}, \Upsilon_{j+1,k}, \Upsilon_{j,k}, \Upsilon_{j-1,k}) \\
\Upsilon_{j,k} &= \left| \frac{p_{j+1,k} - 2p_{j,k} + p_{j-1,k}}{p_{j+1,k} + 2p_{j,k} + p_{j-1,k}} \right| \\
\epsilon_l^{(4)} &= \max(0, \kappa_4 - \epsilon_l^{(2)}),
\end{aligned} \tag{5.20}$$

where typical values of the constants are $\kappa_2 = 1/2$ and $\kappa_4 = 1/32$.

The artificial dissipation terms for the edges with constant η are analogous. They are obtained by replacing ξ with η in (5.16) and (5.17).

As described, this artificial dissipation model parallels that used with the implicit algorithm described in Chap. 4. When used with an explicit multigrid algorithm, it is sometimes modified in the following manner [7]. The spectral radius associated with the ξ direction given in (5.19) is multiplied by $\phi(r)$, which is given by

$$\phi(r_{\eta\xi}) = 1 + r_{\eta\xi}^{\zeta}, \tag{5.21}$$

with

$$r_{\eta\xi} = \frac{\lambda_\eta}{\lambda_\xi}, \tag{5.22}$$

where ζ is typically equal to 2/3. The spectral radius in the η direction λ_η is multiplied by $\phi(r^{-1})$. This increases the amount of numerical dissipation, thus improving the high-frequency damping properties of the scheme and leading to better convergence rates with the multigrid method. This is particularly important in the case of high aspect ratio cells, for example in high Reynolds number boundary layers. In such cases, the ratio λ_η/λ_ξ approximates the cell aspect ratio. With a cell aspect ratio of 1000, for example, ϕ is on the order of 100, and the numerical dissipation in the streamwise direction is greatly increased.

5.3 Iteration to Steady State

5.3.1 Multi-stage Time-Marching Method

The semi-discrete form (5.6) can be written as

$$\frac{\mathrm{d}}{\mathrm{d}t} Q_{j,k} = -\frac{1}{A_{j,k}} \mathcal{L} Q_{j,k}, \tag{5.23}$$

where $\mathcal{L} = \mathcal{L}_{\mathrm{i}} + \mathcal{L}_{\mathrm{ad}} - \mathcal{L}_{\mathrm{v}}$. Here we will concentrate on an explicit multi-stage time-marching method which can be used for steady flows or to solve the nonlinear problem arising at each time step in the dual-time-stepping approach to unsteady flows (see Sect. 4.5.7). In both of these settings there is no benefit to higher-order accuracy in time, and we will consider methods designed specifically for rapid convergence to steady state when used in conjunction with the multigrid method.

The effectiveness of a time-marching method for convergence to a steady state can be assessed in terms of the amplification factor (based on the σ eigenvalues in the terminology of Chap. 2) arising from the λh eigenvalues resulting from a specific spatial discretization. This is discussed further below, but we begin with a more qualitative discussion. When iterations are performed from an arbitrary initial condition to the steady-state solution, we can consider the difference between the initial condition and the steady solution to be an *error* that must be removed. Since the time-marching iterations represent a physical process, one can give a physical interpretation of the path to steady state. The error is removed through two mechanisms associated with the governing PDEs: (1) it convects out of the domain through the boundary, and (2) it dissipates within the domain through both physical and numerical dissipation. If one thinks of the error as being decomposed into modes, then low frequency error modes will typically be eliminated through convection and high frequency modes through dissipation.

A time-marching method with good convergence properties addresses these two mechanisms in the following manner. In order to enable convection of the error through the boundary, the method should be at least second-order accurate, so that the physics of convection is accurately represented, and when combined with a particular spatial discretization, the maximum stable Courant number should be as large as possible. The method should also provide damping of high frequency modes, again in combination with the spatial discretization. The latter property is particularly important in the context of the multigrid method, which will be discussed in Sect. 5.3.3. Finally, the computational cost per time step is also an important consideration.

We will begin by considering a time-marching method for the spatially discretized Euler equations, i.e. applied to the ODE system

$$\frac{\mathrm{d}}{\mathrm{d}t} Q_{j,k} = -\frac{1}{A_{j,k}} (\mathcal{L}_{\mathrm{i}} + \mathcal{L}_{\mathrm{ad}}) Q_{j,k} = -R(Q_{j,k}). \tag{5.24}$$

Consider a multi-stage time marching method in the following form

$$\begin{aligned}
Q_{j,k}^{(0)} &= Q_{j,k}^{(n)} \\
Q_{j,k}^{(m)} &= Q_{j,k}^{(0)} - \alpha_m h R(Q_{j,k}^{(m-1)}), \qquad m = 1, \ldots, q \\
Q_{j,k}^{(n+1)} &= Q_{j,k}^{(q)},
\end{aligned} \tag{5.25}$$

where n is the time index, $h = \Delta t$, q is the number of stages, and the coefficients $\alpha_m, m = 1, \ldots, q$ define the method. The reader should recognize that this is not a general form for explicit Runge-Kutta methods. For example, the classical fourth-order method given in Sect. 2.6 cannot be written in this form. Nevertheless, this form is equivalent to the more general form with respect to homogeneous ODEs and thus enables the design of schemes with tailored convergence properties.

For the purpose of discussing the analysis of such methods we will concentrate on five-stage methods, i.e. $q = 5$. Consider the homogeneous scalar ODE given by

$$\frac{\mathrm{d}u}{\mathrm{d}t} = \lambda u, \tag{5.26}$$

where λ represents an eigenvalue of the linearized semi-discrete system. When applied to this ODE, the method given by (5.25) with $q = 5$ produces the solution

$$u_n = u_0 \sigma^n, \tag{5.27}$$

where u_0 is the initial condition, and σ is given by

$$\sigma = 1 + \beta_1 \lambda h + \beta_2(\lambda h)^2 + \beta_3(\lambda h)^3 + \beta_4(\lambda h)^4 + \beta_5(\lambda h)^5, \tag{5.28}$$

with

$$\begin{aligned}
\beta_1 &= \alpha_5 \\
\beta_2 &= \alpha_5\alpha_4 \\
\beta_3 &= \alpha_5\alpha_4\alpha_3 \\
\beta_4 &= \alpha_5\alpha_4\alpha_3\alpha_2 \\
\beta_5 &= \alpha_5\alpha_4\alpha_3\alpha_2\alpha_1 .
\end{aligned} \tag{5.29}$$

Second-order accuracy is obtained by choosing $\alpha_5 = 1$ and $\alpha_4 = 1/2$, giving $\beta_1 = 1$ and $\beta_2 = 1/2$. This leaves three free parameters that can be chosen from the perspective of optimizing convergence to steady state.

The values $\beta_3 = 1/6$, $\beta_4 = 1/24$, and $\beta_5 = 1/120$ lead to a σ that approximates $\mathrm{e}^{\lambda h}$, which maximizes the order of accuracy of the method, at least for homogeneous ODEs such as (5.26). This is obtained with $\alpha_1 = 1/5$, $\alpha_2 = 1/4$, and $\alpha_3 = 1/3$. Figure 5.3 shows contours of $|\sigma|$ for this method plotted in the complex λh plane. The method has a large region of stability that includes a portion of the imaginary axis.

The convergence rates this method will produce depend upon the specific spatial discretization and the time step. To examine this, consider the linear convection equation

$$\frac{\partial u}{\partial t} + a\frac{\partial u}{\partial x} = 0, \tag{5.30}$$

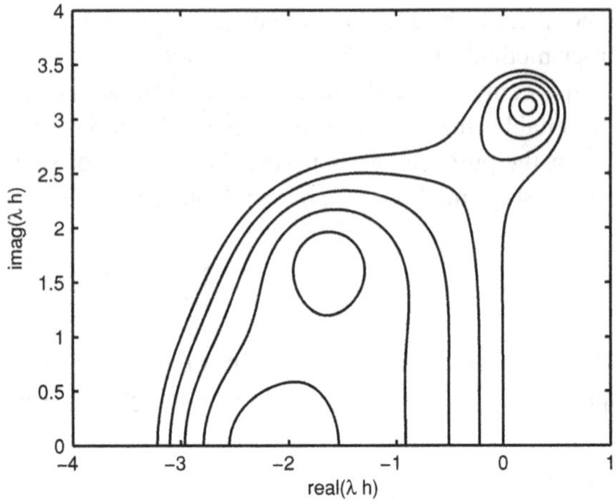

Fig. 5.3 Contours of $|\sigma|$ for the five-stage time-marching method with $\beta_3 = 1/6$, $\beta_4 = 1/24$, and $\beta_5 = 1/120$. Contours shown have $|\sigma|$ equal to 1, 0.8, 0.6, 0.4, and 0.2

with $a > 0$ and periodic boundary conditions. Apply second-order centered differences with fourth-difference artificial dissipation to approximate the spatial derivative term:

$$-a\delta_x u = -\frac{a}{\Delta x}\left[\frac{u_{j+1} - u_{j-1}}{2} + \kappa_4(u_{j-2} - 4u_{j-1} + 6u_j - 4u_{j+1} + u_{j+2})\right]. \tag{5.31}$$

Since the boundary conditions are periodic, Fourier analysis can be used to obtain the λ eigenvalues of the resulting semi-discrete form. They are given by

$$\lambda_m = -\frac{a}{\Delta x}\left\{i \sin\left(\frac{2\pi m}{M}\right) + 4\kappa_4\left[1 - \cos\left(\frac{2\pi m}{M}\right)\right]^2\right\}, \quad m = 0 \ldots M-1, \tag{5.32}$$

where M corresponds to the number of nodes in the mesh. Multiplying by the time step gives

$$\lambda_m h = -C_n\left\{i \sin\left(\frac{2\pi m}{M}\right) + 4\kappa_4\left[1 - \cos\left(\frac{2\pi m}{M}\right)\right]^2\right\}, \quad m = 0 \ldots M-1, \tag{5.33}$$

where $C_n = ah/\Delta x$ is the Courant number.

The λh values given by (5.33) are plotted in Fig. 5.4 for $M = 40$, $\kappa_4 = 1/32$, and $C_n = 2.5$ together with the $|\sigma|$ contours arising from the five-stage scheme (5.25) with $\alpha_1 = 1/5$, $\alpha_2 = 1/4$, and $\alpha_3 = 1/3$. Figure 5.5 plots $|\sigma(\lambda_m h)|$ vs. $\kappa\Delta x$ for $0 \leq \kappa\Delta x \leq \pi$, where $\kappa\Delta x = 2\pi m/M$. This plot shows poor damping for

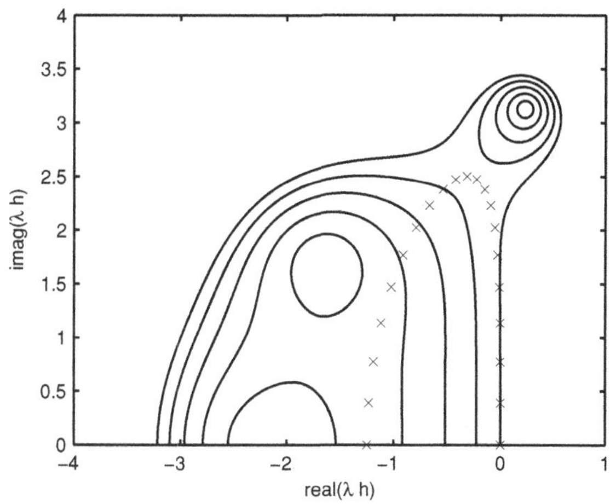

Fig. 5.4 Plot of λh values given by (5.33) for $M = 40$, $\kappa_4 = 1/32$, and $C_n = 2.5$ with contours of $|\sigma|$ for the five-stage time-marching method with $\alpha_1 = 1/5$, $\alpha_2 = 1/4$, and $\alpha_3 = 1/3$. Contours shown have $|\sigma|$ equal to 1, 0.8, 0.6, 0.4, and 0.2

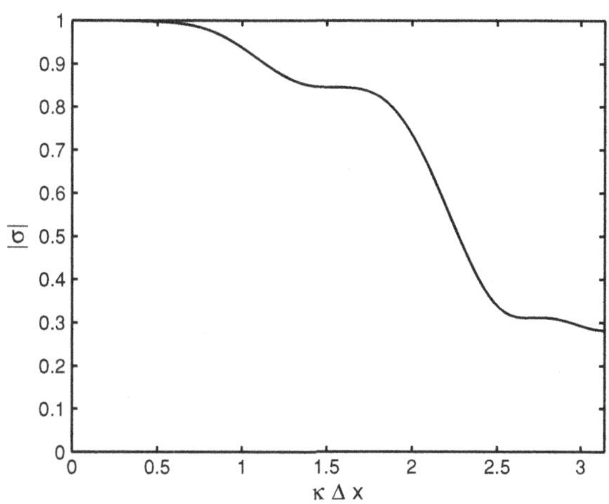

Fig. 5.5 Plot of $|\sigma|$ values vs. $\kappa \Delta x$ for the spatial operator given by (5.31) with $C_n = 2.5$, $\kappa_4 = 1/32$, and the five-stage time-marching method with $\alpha_1 = 1/5$, $\alpha_2 = 1/4$, and $\alpha_3 = 1/3$

low wavenumbers and good damping at high wavenumbers. As we shall see later, this provides a smoothing property suitable for use with the multigrid method. It is important to recognize that this model problem includes only the mechanism of damping within the domain. With periodic boundary conditions, the error cannot convect out of the domain, so this mechanism is not represented. Therefore, the Courant number is also an important quantity to be aware of. Although the effect is not seen in the present analysis, a higher stable Courant number translates into a larger time step, which enables the error to convect out through the outer boundary of the domain in fewer time steps.

Through careful selection of the free parameters, α_1, α_2, and α_3, a multi-stage method can be designed for fast convergence when used in conjunction with a specific spatial discretization. For example, consider the choice $\alpha_1 = 1/4$, $\alpha_2 = 1/6$, and $\alpha_3 = 3/8$, which maximizes the stable region on the imaginary axis (see Van der

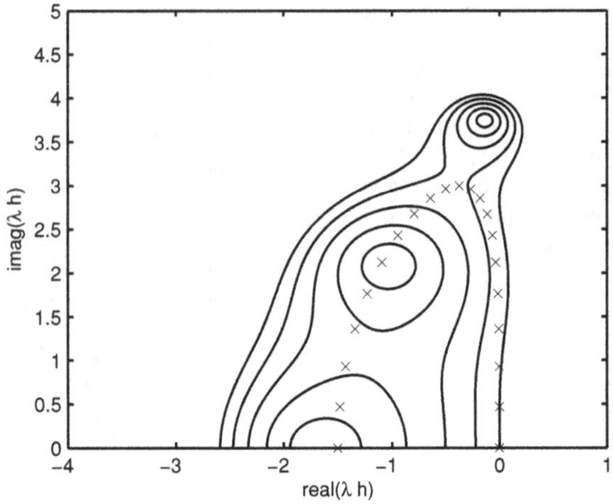

Fig. 5.6 Plot of λh values given by (5.33) for $M = 40$, $\kappa_4 = 1/32$, and $C_n = 3$ with contours of $|\sigma|$ for the five-stage time-marching method with $\alpha_1 = 1/4$, $\alpha_2 = 1/6$, and $\alpha_3 = 3/8$

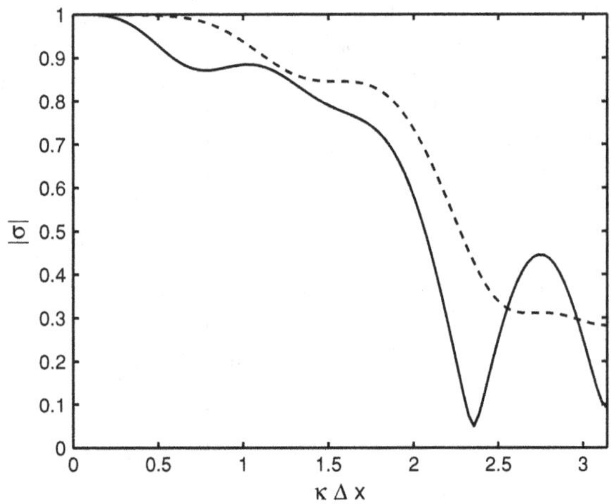

Fig. 5.7 Plot of $|\sigma|$ values vs. $\kappa\Delta x$ for the spatial operator given by (5.31) with $C_n = 3$, $\kappa_4 = 1/32$, and the five-stage time-marching method with $\alpha_1 = 1/4$, $\alpha_2 = 1/6$, and $\alpha_3 = 3/8$ (solid line). The *dashed line* shows the results with $C_n = 2.5$ and $\alpha_1 = 1/5$, $\alpha_2 = 1/4$, and $\alpha_3 = 1/3$

Houwen [9]). The associated plots are shown in Figs. 5.6 and 5.7 with a Courant number of 3. The improvement in damping properties is small, but the higher Courant number enables the error to be propagated to the outer boundary more rapidly. This particular choice of α coefficients is intended for use with a spatial discretization that combines centered differencing (or an equivalent finite-volume method) with artificial dissipation. One can also design multi-stage schemes specifically for upwind schemes.

One must be aware of the limitations of such scalar Fourier analysis in this context. It provides a useful guide for the design of multi-stage schemes, but, since it does not account for systems of PDEs, multidimensionality, or the effect of boundaries, the performance of such schemes when applied to the Euler equations must be assessed through more sophisticated theory or numerical experiment.

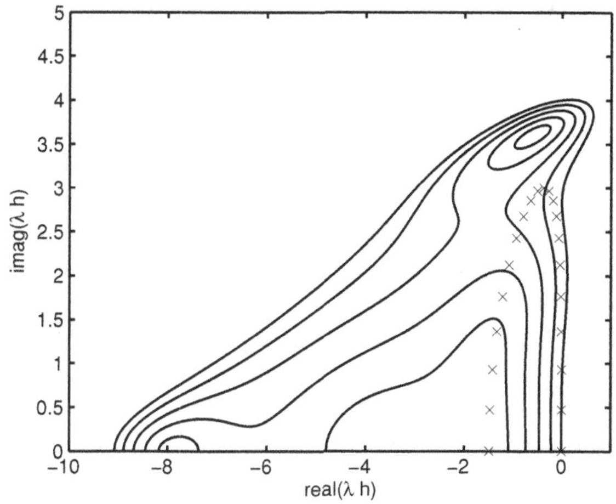

Fig. 5.8 Plot of λh values given by (5.33) for $M = 40$, $\kappa_4 = 1/32$, and $C_n = 3$ with contours of $|\sigma|$ for the five-stage time-marching method with $\alpha_1 = 1/4, \alpha_2 = 1/6$, and $\alpha_3 = 3/8$ with the artificial dissipation computed only on stages 1, 3, and 5

A further generalization of (5.25) can be introduced if the distinct components of $R(Q)$, for example $\mathcal{L}_i Q$ and $\mathcal{L}_{ad} Q$, are handled differently by the multi-stage method. Consider a scheme where at stage m the residual term $R(Q_{j,k}^{(m-1)})$ in (5.25) is replaced by

$$R^{(m-1)} = \frac{1}{A}\left(\mathcal{L}_i Q^{(m-1)} + \sum_{p=0}^{m-1} \gamma_{mp} \mathcal{L}_{ad} Q^{(p)}\right). \tag{5.34}$$

The γ_{mp} coefficients can be chosen such that the artificial dissipation operator is evaluated only at certain stages, thus reducing the computational effort per time step. The following values lead to a method in which the artificial dissipation is evaluated at the first, third, and fifth stages:

$$\begin{aligned}
&\gamma_{10} = 1\\
&\gamma_{20} = 1, \quad \gamma_{21} = 0\\
&\gamma_{30} = 1 - \Gamma_3, \quad \gamma_{31} = 0, \quad \gamma_{32} = \Gamma_3\\
&\gamma_{40} = 1 - \Gamma_3, \quad \gamma_{41} = 0, \quad \gamma_{42} = \Gamma_3, \quad \gamma_{43} = 0\\
&\gamma_{50} = (1 - \Gamma_3)(1 - \Gamma_5), \quad \gamma_{51} = 0, \quad \gamma_{52} = \Gamma_3(1 - \Gamma_5), \quad \gamma_{53} = 0, \quad \gamma_{54} = \Gamma_5.
\end{aligned} \tag{5.35}$$

Note that the coefficients sum to unity at each stage. With $\Gamma_3 = 0.56$ and $\Gamma_5 = 0.44$, the results shown in Figs. 5.8 and 5.9 are obtained for the linear convection equation. This method retains the favourable damping properties of the previous method while reducing the computational cost per time step, thereby reducing the overall cost to achieve a converged solution.

The above multi-stage method is also appropriate for the numerical solution of the Navier-Stokes equations. In this case, the residual includes the contribution from

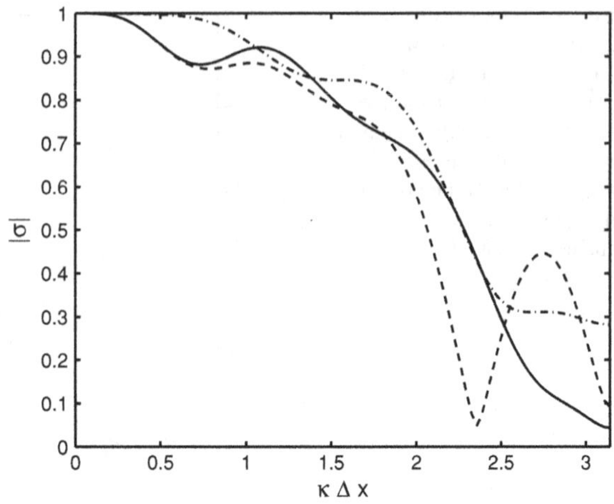

Fig. 5.9 Plot of $|\sigma|$ values vs. $\kappa\Delta x$ for the spatial operator given by (5.31) with $C_n = 3$, $\kappa_4 = 1/32$, and the five-stage time-marching method with $\alpha_1 = 1/4$, $\alpha_2 = 1/6$, and $\alpha_3 = 3/8$ with the artificial dissipation computed only on stages 1, 3, and 5 (*solid line*). The *dashed line* shows the results with the artificial dissipation computed at every stage, and the *dash-dot line* shows the results with $C_n = 2.5$ and $\alpha_1 = 1/5$, $\alpha_2 = 1/4$, and $\alpha_3 = 1/3$

the viscous and heat conduction terms, $\mathcal{L}_v$. The residual can be computed at each stage as follows:

$$R^{(m-1)} = \frac{1}{A}\left(\mathcal{L}_i Q^{(m-1)} - \mathcal{L}_v Q^{(0)} + \sum_{p=0}^{m-1} \gamma_{mp}\mathcal{L}_{ad} Q^{(p)}\right). \tag{5.36}$$

The viscous terms are evaluated at the first stage only, thereby minimizing the additional cost per time step.

Local Time Stepping. Use of a local time step specific to each grid cell is important to improve the convergence rate of an explicit algorithm for steady flows. In order to understand why, consider first the use of a constant time step. For example, for the one-dimensional Euler equations we have

$$\Delta t \leq \frac{\Delta x}{|u| + a}(C_n)_{max}, \tag{5.37}$$

where $|u| + a$ is the largest eigenvalue of the flux Jacobian, and $(C_n)_{max}$ is the maximum Courant number for stability of the particular combination of spatial discretization and time-marching method, as determined by Fourier analysis, for example (bearing in mind that Fourier analysis provides a necessary condition for stability but not a sufficient one). The stability requirement resulting from the conditional stability associated with explicit schemes will dictate that the time step be determined based on the grid cell with the smallest value of $\Delta x/(|u|+a)$. Typically the variation in mesh spacing far exceeds the variation in the maximum wave speed; hence the time step is often limited by the size of the smallest cells in the mesh. If the smallest cells are several orders of magnitude smaller than the largest cells, then this time step will be much smaller than the optimal time step for the larger cells.

We can assign a physical meaning to the Courant number. It is the distance travelled by the fastest wave in one time step expressed in terms of the mesh spacing. For example, with a Courant number of 3, the fastest wave travels a distance $3\Delta x$ in one time step. However, if the time step is determined by a very small cell, then the effective Courant number at a large cell is very small, and it will take many time steps for a disturbance to propagate through the large cell.

On a mesh with a wide variation in mesh spacing, much faster convergence to steady state can be achieved by using a time step at each cell that gives the desired value of the Courant number for that cell. For example, in our one-dimensional example the local time step is computed from

$$(\Delta t)_j = \frac{(\Delta x)_j}{(|u| + a)_j} C_{\text{n}}, \tag{5.38}$$

where C_{n} is the desired (optimal) Courant number. The use of such a local time step destroys time accuracy but has no impact on the converged steady solution.

For the one-dimensional Euler equations, the definition of the local time step (5.38) is a relatively straightforward matter. Extension to multidimensions and to the Navier-Stokes equations is not straightforward, and a number of approximations are typically made. In order to present some of the issues, we will consider the convection-diffusion equation as a model problem:

$$\frac{\partial u}{\partial t} + a\frac{\partial u}{\partial x} = \nu\frac{\partial^2 u}{\partial x^2}. \tag{5.39}$$

With periodic boundary conditions and second-order centered-difference approximations to both the first and the second spatial derivatives on a mesh with M nodes, the eigenvalues of the semi-discrete operator matrix are, from Fourier analysis:

$$\lambda_m = -\frac{a}{\Delta x} i \sin\left(\frac{2\pi m}{M}\right) - \frac{4\nu}{\Delta x^2}\sin^2\left(\frac{\pi m}{M}\right), \quad m = 0, \ldots, M-1, \tag{5.40}$$

where $\Delta x = 2\pi/M$. The imaginary part of the eigenvalue is associated with the convective term, the real part with the diffusive term.

Let us consider the solution of this semi-discrete system using the five-stage time-marching method described previously with $\alpha_1 = 1/4$, $\alpha_2 = 1/6$, and $\alpha_3 = 3/8$. From Fig. 5.6 we see that this method is stable for imaginary eigenvalues up to 4 and for negative real eigenvalues up to -2.59. We will attempt to define a local time step based solely on this information about the time-marching method. Multiplying the above eigenvalues by h gives

$$\lambda_m h = -C_{\text{n}} i \sin\left(\frac{2\pi m}{M}\right) - 4V_{\text{n}}\sin^2\left(\frac{\pi m}{M}\right), \quad m = 0, \ldots, M-1, \tag{5.41}$$

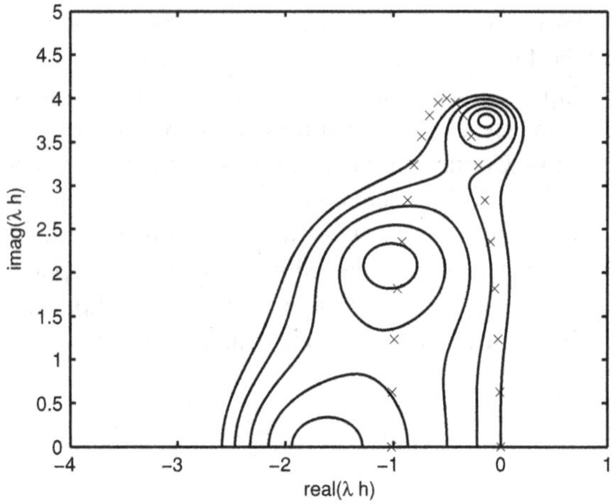

Fig. 5.10 Plot of λh values given by (5.33) for $M = 40$, $\kappa_4 = 1/32$, and $C_n = 3$ with contours of $|\sigma|$ for the five-stage time-marching method with $\alpha_1 = 1/4, \alpha_2 = 1/6$, and $\alpha_3 = 3/8$. Time step based on minimum of h_c and h_d

where $V_n = \nu h/\Delta x^2$ is sometimes referred to as the von Neumann number. Based on the above properties of the time-marching method, we require for stability:

$$C_n = \frac{ah}{\Delta x} \leq 4$$
$$V_n = \frac{\nu h}{\Delta x^2} \leq \frac{2.59}{4}. \tag{5.42}$$

Based on the first criterion, one can define a convective time step limit as

$$h_c \leq \frac{4\Delta x}{a}, \tag{5.43}$$

while the second criterion gives the diffusive time step limit as

$$h_d \leq \frac{2.59\Delta x^2}{4\nu}. \tag{5.44}$$

It is tempting, therefore, to choose the time step as the minimum of h_c and h_d, which ensures that the imaginary part of all eigenvalues is less than 4 and the negative real part is less than 2.5. However, consider an example with $a = 1$, $\nu = 0.01$ and $M = 40$. The resulting spectrum is displayed in Fig. 5.10 along with the $|\sigma|$ contours of the time-marching method. Some eigenvalues lie outside the stable region; hence this time step definition is not adequate to ensure stability.

A more conservative time step definition is obtained from

$$\frac{1}{h} = \frac{1}{h_c} + \frac{1}{h_d}. \tag{5.45}$$

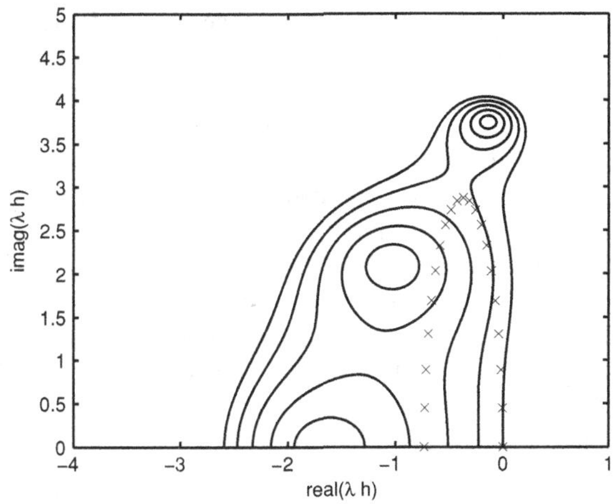

Fig. 5.11 Plot of λh values given by (5.33) for $M = 40$, $\kappa_4 = 1/32$, and $C_n = 3$ with contours of $|\sigma|$ for the five-stage time-marching method with $\alpha_1 = 1/4, \alpha_2 = 1/6$, and $\alpha_3 = 3/8$. Time step based on (5.45)

With this choice, the time step is less than the minimum of h_c and h_d. For the above example, the λh values plotted in Fig. 5.11 are obtained. All of the eigenvalues lie well within the stable region of the time-marching method.

Based on approximations such as this, various local time stepping strategies have been developed for explicit multi-stage time-marching methods with the goal of providing robust and rapid convergence. One such approach, which is based on (5.45), is given by Swanson and Turkel [7] as follows:

$$h = \frac{N_i A}{\lambda_C + \lambda_D}, \tag{5.46}$$

where

$$\begin{aligned} \lambda_C &= \lambda_\xi + \lambda_\eta \\ \lambda_D &= (\lambda_D)_\xi + (\lambda_D)_\eta + (\lambda_D)_{\xi\eta}, \end{aligned} \tag{5.47}$$

with

$$\begin{aligned} (\lambda_D)_\xi &= \frac{\gamma\mu}{Re\rho Pr} A^{-1}(x_\eta^2 + y_\eta^2) \\ (\lambda_D)_\eta &= \frac{\gamma\mu}{Re\rho Pr} A^{-1}(x_\xi^2 + y_\xi^2) \\ (\lambda_D)_{\xi\eta} &= \frac{\mu}{Re\rho} A^{-1}\left[-\frac{7}{3}(y_\eta y_\xi + x_\xi x_\eta) + \frac{1}{3}\sqrt{(x_\eta^2 + y_\eta^2)(x_\xi^2 + y_\xi^2)}\right]. \end{aligned} \tag{5.48}$$

The quantity N_i is the stability bound on pure imaginary eigenvalues associated with the time-marching method used. The assumption made is that the maximum negative real eigenvalue is of a similar magnitude. The cell area is denoted by A, and the terms λ_ξ and λ_η are defined as in (5.19). For the cell in question, λ_ξ is obtained by averaging the values obtained from the two edges of constant ξ, while λ_η is obtained by averaging the values obtained from the two edges of constant η. The diffusive terms $(\lambda_D)_\xi$, $(\lambda_D)_\eta$, and $(\lambda_D)_{\xi\eta}$ are approximations to the spectral radii of the respective viscous flux Jacobians. The metric terms appearing in these terms are also calculated based on undivided differences for the appropriate edges and then averaging to get a value for the cell. For example, y_η is obtained by averaging Δy for opposing edges of constant ξ, and the other terms are obtained similarly.

Given the various approximations made in determining the local time step for the Navier-Stokes equations in multidimensions, it is typical to include a factor in the time step definition that is determined to be effective, i.e. both reliable and efficient, through numerical experimentation. The use of a local time step enables fast convergence of an explicit method on a mesh with a large variation in mesh spacing. However, it does not address the slow convergence of explicit methods resulting from grid cells with high aspect ratios.

5.3.2 Implicit Residual Smoothing

Implicit residual smoothing is a convergence acceleration technique that enables a substantial increase in the Courant number, thus speeding up the propagation of disturbances to the outer boundary. First we define a residual that incorporates the local time step:

$$\tilde{R}_{j,k}^{(m-1)} = \frac{(\Delta t)_{j,k}}{A_{j,k}} \left(\mathcal{L}_{\mathrm{i}} Q_{j,k}^{(m-1)} - \mathcal{L}_{\mathrm{v}} Q_{j,k}^{(0)} + \sum_{p=0}^{m-1} \gamma_{mp} \mathcal{L}_{\mathrm{ad}} Q_{j,k}^{(p)} \right). \tag{5.49}$$

A smoothed residual $\bar{R}_{j,k}^{(m-1)}$ is found from the following:

$$(1 - \beta_\xi \nabla_\xi \Delta_\xi)(1 - \beta_\eta \nabla_\eta \Delta_\eta) \bar{R}_{j,k}^{(m-1)} = \tilde{R}_{j,k}^{(m-1)} \tag{5.50}$$

and replaces the term $hR(Q_{j,k}^{(m-1)})$ in (5.25). As in Sect. 5.2.2, Δ_ξ and ∇_ξ represent undivided differences in the ξ direction, and Δ_η and ∇_η are the corresponding operators in the η direction. The smoothing coefficients β_ξ and β_η are discussed below. The operator in the ξ direction can be rewritten as

$$(1 - \beta_\xi \nabla_\xi \Delta_\xi) \bar{R}_{j,k}^{(m-1)} = \left[-\beta_\xi \bar{R}_{j-1,k}^{(m-1)} + (1 + 2\beta_\xi) \bar{R}_{j,k}^{(m-1)} - \beta_\xi \bar{R}_{j+1,k}^{(m-1)} \right]. \tag{5.51}$$

The residuals of the individual equations, i.e. mass, x and y-momentum, and energy, are smoothed separately. Hence in two dimensions implicit residual smoothing requires the solution of two scalar tridiagonal systems per stage of the multi-stage time-stepping scheme. This adds considerably to the computational cost per time step.

In order to understand and analyze implicit residual smoothing, we return to the linear convection equation with periodic boundary conditions discretized using the operator given in (5.31). In a one-dimensional scalar problem, the implicit residual smoothing operator is given by

$$B_p(M : -\beta, 1 + 2\beta, -\beta)\bar{R} = R, \tag{5.52}$$

or

$$\bar{R} = [B_p(M : -\beta, 1 + 2\beta, -\beta)]^{-1} R, \tag{5.53}$$

where we use the notation for the banded periodic matrix $B_p(M : a, b, c)$ given in (2.33). Hence we can obtain the eigenvalues of the system with implicit residual smoothing by dividing those given in (5.33) by the eigenvalues of $B_p(M : -\beta, 1 + 2\beta, -\beta)$,[1] leading to

$$\lambda_m h = -C_n \frac{i \sin\left(\frac{2\pi m}{M}\right) + 4\kappa_4 \left[1 - \cos\left(\frac{2\pi m}{M}\right)\right]^2}{1 + 4\beta \sin^2\left(\frac{\pi m}{M}\right)}, \quad m = 0 \ldots M - 1. \tag{5.54}$$

For the problem studied previously, with $M = 40$, $C_n = 3$, and $\kappa = 1/32$ coupled with a smoothing coefficient of $\beta = 0.6$, the eigenvalues λh are displayed in Fig. 5.12. There are two primary observations to be made. First, the magnitude of the eigenvalues has generally been reduced as a result of the implicit residual smoothing. This means that a larger Courant number can be used while remaining within the stability bounds of a given time-marching method. Second, the eigenvalues associated with small m, which are those at the origin and just above and below, are affected the least by the residual smoothing. These eigenvalues correspond to well resolved modes, i.e. low frequency modes, which are those that convect out through the boundary. Hence the residual smoothing has little effect on the manner in which these modes are propagated.

Figure 5.13 shows these eigenvalues superimposed on the $|\sigma|$ contours of the five-stage method with dissipation evaluated on the first, third, and fifth stages. As a result of the implicit residual smoothing, a Courant number of 7 can be used while remaining in the stable region. Consequently, disturbances will propagate to the outer boundary in fewer time steps than without residual smoothing (where the Courant number is 3). Figure 5.14 shows that the damping properties are similar to those obtained without residual smoothing, so the primary benefit is the higher Courant number. It is important to recognize that the use of implicit residual smoothing entails

[1] as a result of the properties of circulant matrices

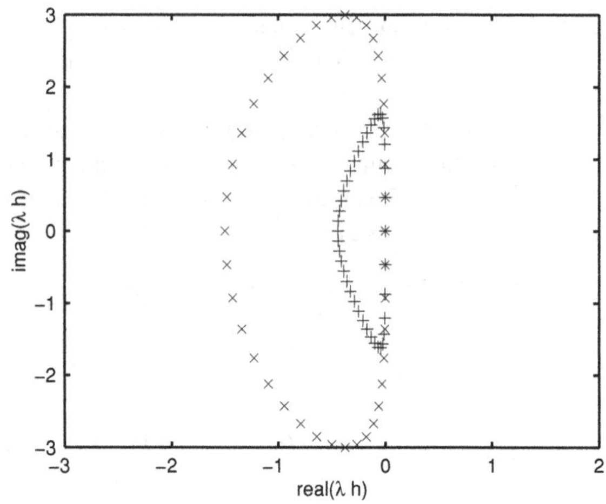

Fig. 5.12 Plot of λh values given by (5.33) for $M = 40$, $\kappa_4 = 1/32$, and $C_n = 3$ without implicit residual smoothing (x) and with implicit residual smoothing (+) with $\beta = 0.6$

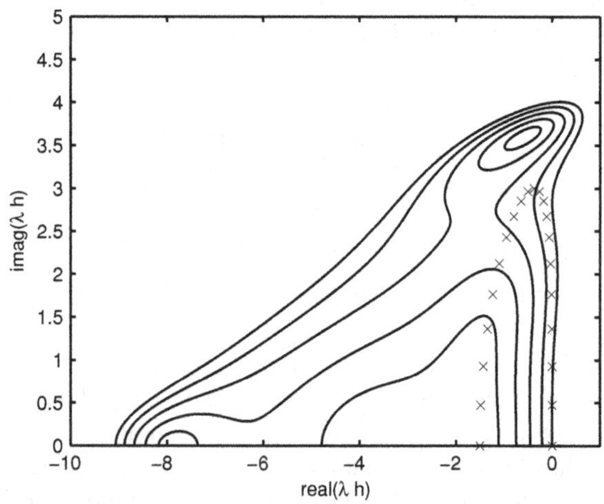

Fig. 5.13 Plot of λh values given by (5.33) for $M = 40$, $\kappa_4 = 1/32$, and $C_n = 7$ with implicit residual smoothing with $\beta = 0.6$ and contours of $|\sigma|$ for the five-stage time-marching method with $\alpha_1 = 1/4$, $\alpha_2 = 1/6$, and $\alpha_3 = 3/8$ with the artificial dissipation computed only on stages 1, 3, and 5

a significant computational expense per time step that must be weighed against the reduced number of time steps to steady state associated with the increased Courant number.

The main purpose of implicit residual smoothing is to enable the use of a larger time step, or Courant number. Typically the Courant number limit is increased by a factor of two to three. This enables disturbances to propagate more rapidly to the domain boundary without compromising damping properties, as shown in Fig. 5.14. The maximum stable Courant number continues to increase as β is increased. However, at some point this does not lead to faster convergence, and there is an optimum value of β. The reason is that the implicit residual smoothing eliminates time accuracy and therefore interferes with the physics of convection and hence the propagation of error to the outer boundary. In Fig. 5.12 we saw that the smaller eigenvalues are not

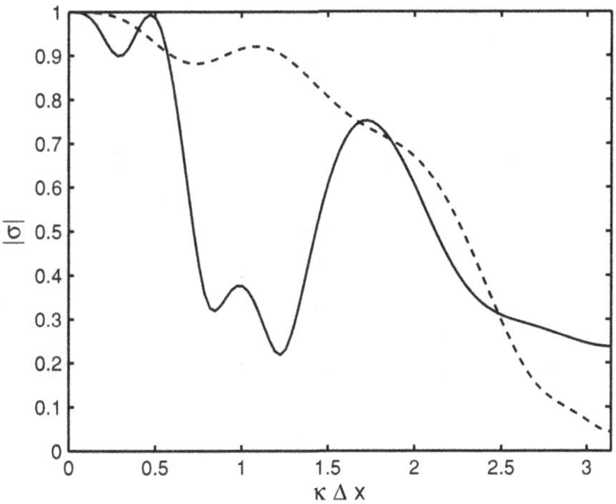

Fig. 5.14 Plot of $|\sigma|$ values vs. $\kappa\Delta x$ for the spatial operator given by 5.31 with $C_n = 7$ with implicit residual smoothing ($\beta = 0.6$), $\kappa_4 = 1/32$, and the five-stage time-marching method with $\alpha_1 = 1/4$, $\alpha_2 = 1/6$, and $\alpha_3 = 3/8$ with the artificial dissipation computed only on stages 1, 3, and 5 (solid line). The *dashed line* shows the results without implicit residual smoothing with $C_n = 3$

greatly affected by the smoothing with $\beta = 0.6$. As β is increased, these eigenvalues begin to deviate more from their values without implicit residual smoothing. Hence there is a compromise between a large Courant number and accurate representation of the convection process for low frequency modes.

Based on one- and two-dimensional stability analysis as well as numerical experiments, Swanson and Turkel [7] developed the following formulas for β_ξ and β_η:

$$\beta_\xi = \max\left\{\frac{1}{4}\left[\left(\frac{N}{N^*}\frac{1}{1+\psi r_{\eta\xi}}\right)^2 - 1\right], 0\right\}$$

$$\beta_\eta = \max\left\{\frac{1}{4}\left[\left(\frac{N}{N^*}\frac{1}{1+\psi r_{\eta\xi}^{-1}}\right)^2 - 1\right], 0\right\}. \qquad (5.55)$$

Here N^* is the Courant number for the unsmoothed scheme, while N is the Courant number for the smoothed scheme, so N/N^* typically takes a value between 2 and 3. The ratio of inviscid spectral radii was defined in (5.22), and ψ is a user-defined parameter generally between 0.125 and 0.25.

5.3.3 *The Multigrid Method*

The multigrid method systematically uses sets of coarser grids to accelerate the convergence of iterative schemes. It can be applied to any iterative method that displays a smoothing property, i.e. it preferentially damps high-frequency error modes. For explicit iterative methods, multigrid is critical to obtaining fast convergence to steady-state for stiff problems.

Multigrid theory is well developed for elliptic problems, such as the steady diffusion equation. For such problems, there is a correlation between the eigenvalues and the spatial frequencies of the associated eigenvectors. For example, for the diffusion equation, the eigenvalues of the semi-discrete operator matrix resulting from a second-order centered-difference discretization are all real and negative (see Sect. 2.3.4). The eigenvectors associated with the eigenvalues with small magnitudes have low spatial frequencies, while those corresponding to eigenvalues with large magnitudes have high frequencies. This means that in the exact solution of the semi-discrete ODE system (see Sect. 2.3.3) the high frequency components in the transient solution are rapidly damped, while the low frequency components are slowly damped. This is a fundamental property of a diffusive system that is retained after discretizing in space.

Given this correlation between eigenvalues with large magnitudes and high space frequencies, it is a natural property of several iterative methods (such as the Gauss-Seidel relaxation method) to reduce error components corresponding to high spatial frequencies more effectively than those corresponding to low spatial frequencies. Moreover, iterative methods can be specifically designed to have this property, such as the Richardson method described in Lomax et al. [10]. The multigrid method exploits this property by systematically using coarser grids to target the removal of specific components of the error. For example, high frequency error components are rapidly damped on the initial grid, whose density is determined by accuracy considerations. Hence the error is *smoothed* on that mesh. The low frequency error components can be represented on a coarser mesh on which some of them appear as high frequencies, where the frequency is relative to the mesh spacing, and are thus more rapidly damped.

To make this clearer, consider the range of wavenumbers that are representable on a mesh with spacing Δx_f, which are given by $0 \leq \kappa \Delta x_f \leq \pi$. If the mesh spacing is increased by a factor of two ($\Delta x_c = 2\Delta x_f$), then the wavenumber range $\pi/2 \leq \kappa \Delta x_f \leq \pi$ on the original mesh cannot be represented on the coarse mesh. However, the range of error modes with $0 \leq \kappa \Delta x_f \leq \pi/2$ have their value of $\kappa \Delta x$ doubled. Those error modes in the wavenumber range $\pi/4 \leq \kappa \Delta x_f \leq \pi/2$ on the fine mesh, which are poorly damped compared to those in the high wavenumber range, appear in the wavenumber range $\pi/2 \leq \kappa \Delta x_c \leq \pi$, which are well damped on the coarse mesh. This can be repeated with successively coarser meshes until the mesh is so coarse that the problem can be affordably solved directly rather than iteratively, such that on that mesh all error modes are damped. This is essentially how the multigrid method works for a linear diffusion problem. See Chap. 10 of [10] for a more detailed description.

Here we are interested in the application of the multigrid method to the discretized Euler and Navier-Stokes equations, which introduces two important differences in comparison to the diffusion equation. First, the Euler and Navier-Stokes equations are nonlinear, which means that the *full approximation storage* approach in which both the residual and the solution must be transferred from the fine to the coarse mesh must be used. Second, in the diffusion problem with Dirichlet boundary conditions, the only mechanism available to remove error modes is diffusion within the domain.

When the Euler and Navier-Stokes equations are solved, error is also propagated through the outer boundary of the domain. This mechanism is primarily associated with low frequency error modes, for which the spatial discretization is relatively accurate. Since such modes are typically poorly damped, this is an important mechanism for their removal. For example, referring to Figs. 5.9 and 5.14, we see that our discretization of the linear convection equation, which includes artificial dissipation, shows preferential damping of high frequencies, i.e. a smoothing property, with the particular time-marching method used.

The analysis reflected in Figs. 5.9 and 5.14 does not include the mechanism of error removal by convection through the boundary. In Sects. 5.3.1 and 5.3.2, we accounted for this by designing schemes to permit as large a Courant number as possible. The multigrid method also exploits this mechanism of error removal. The low frequency error modes for which propagation through the boundary is important are well represented on the coarser mesh. Since the mesh spacing is doubled on the coarse mesh, maintaining a constant Courant number will lead to a doubling of the time step, enabling disturbances to propagate to the outer boundary in roughly half as many time steps. Therefore, when applied to the Euler and Navier-Stokes equations, the multigrid method enhances the convergence rate both through accelerating the damping of error modes within the domain and through accelerating the removal of error through the outer boundary.

We now present the implementation of the multigrid method in conjunction with the cell-centered finite-volume scheme and multi-stage time-marching method described in this chapter. The system of ODEs resulting from the spatial discretization is

$$\frac{\mathrm{d}}{\mathrm{d}t}Q_{j,k} = -\frac{1}{A_{j,k}}\mathcal{L}Q_{j,k} = -R_{j,k}. \tag{5.56}$$

A sequence of grids can be created by successively removing every second grid line in each coordinate direction from the finest grid. The coarse grid cell is then the agglomeration of four fine grid cells sharing a common grid node. If the number of fine grid cells in each coordinate direction is even, then all cells can be merged. Typically sequences of three to five meshes are used. For a five-mesh sequence, the finest mesh should have a number of cells in each direction that is a multiple of 16 in order that the second coarsest mesh have an even number of cells in each direction.

We will now describe a two-grid process that can readily be extended to an arbitrary number of grids, since the process is recursive. We first complete one or more iterations of the five-stage time-marching method with implicit residual smoothing described previously to obtain Q_h. This is followed by an additional computation of the full residual based on the updated solution, including the convective, viscous, and artificial dissipation contributions.

The next step is to transfer the residual and the solution from the fine to the coarse mesh, a process known as *restriction*. Consider the residual first. The term $\mathcal{L}Q_{j,k}$ is the net flux out of cell j, k. In order to transfer the residual to the coarse mesh in a conservative manner, the net flux out of the coarse grid cell should be equal to the net flux out of the four fine grid cells that were merged to form the coarse grid cell. This is achieved simply by summing the flux of each of the four fine grid cells, since internal fluxes will cancel, giving

$$I_h^{2h} R_h = \frac{1}{A_{2h}} \sum_{p=1}^{4} A_h R_h, \tag{5.57}$$

where the subscripts h and $2h$ denote the fine and coarse grids, respectively, and I_h^{2h} is the restriction operator.

An analogous conservative approach is taken to restrict the solution Q. The amount of a conserved quantity, such as mass, momentum, or energy, in the coarse grid cell should be equal to the sum of the amount of that conserved quantity in the constituent fine grid cells. Since Q represents the conserved quantities per unit volume in a given cell, it must be multiplied by the cell area to give the total amount of the conserved quantity in the cell (noting that in two dimensions the conserved quantities are per unit depth). Hence the formula for restricting the solution to the coarse mesh is

$$Q_{2h}^{(0)} = I_h^{2h} Q_h = \frac{1}{A_{2h}} \sum_{p=1}^{4} A_h Q_h, \tag{5.58}$$

where $Q_{2h}^{(0)}$ is the solution used to initiate the multi-stage method on the coarse mesh (see (5.25)).

Now we are ready to solve a problem on the coarse mesh. It is important to recognize that it is not our goal to find the solution to the governing equations on the coarse mesh. The purpose of solving on the coarse mesh is to provide a correction to the solution that will reduce the residual on the fine mesh. To this end, a forcing term P_{2h} is introduced into the ODE solved on the coarse mesh as follows [7]:

$$\frac{\mathrm{d}}{\mathrm{d}t} Q_{2h} = -[R_{2h}(Q_{2h}) + P_{2h}], \tag{5.59}$$

where R_{2h} is the residual computed by applying the spatial discretization on the coarse mesh. The forcing term is

$$P_{2h} = I_h^{2h} R_h - R_{2h}(Q_{2h}^{(0)}), \tag{5.60}$$

which is the difference between the restricted residual and the coarse grid residual computed based on the restricted solution. If we were to drive the coarse mesh problem (5.59) to convergence, we would drive to zero

$$R_{2h}(Q_{2h}) + P_{2h} = R_{2h}(Q_{2h}) - R_{2h}(Q_{2h}^{(0)}) + I_h^{2h} R_h. \tag{5.61}$$

Thus we would obtain the change in the solution on the coarse mesh $(Q_{2h} - Q_{2h}^{(0)})$ that produces a change in the coarse mesh residual $(R_{2h}(Q_{2h}) - R_{2h}(Q_{2h}^{(0)}))$ that offsets the residual restricted from the fine mesh $(I_h^{2h} R_h)$, which is the purpose of the coarse grid correction.

Let us examine the forcing term in more detail. At the first stage of the multi-stage method on the coarse mesh the residual is

$$-[R_{2h}(Q_{2h}^{(0)}) + P_{2h}] = -[R_{2h}(Q_{2h}^{(0)}) + I_h^{2h} R_h - R_{2h}(Q_{2h}^{(0)})] = -I_h^{2h} R_h, \tag{5.62}$$

which is simply the residual restricted from the fine mesh. This means that once the solution on the fine mesh has converged, the coarse mesh calculation will produce no correction, which is appropriate. This provides a useful test when debugging a multigrid algorithm. One can compute the converged solution on the fine mesh using the basic algorithm without multigrid and use this as the initial condition for the multigrid algorithm. Quite a few possible errors can reveal themselves if the coarse mesh correction is nonzero. For example, it is important to enforce the boundary conditions on the coarse mesh before computing the term $R_{2h}(Q_{2h}^{(0)})$ in the forcing function P_{2h}. Otherwise, when they are enforced during the first stage of the multi-stage method, the value of $R_{2h}(Q_{2h}^{(0)})$ will not cancel with the same term in P_{2h}, and a nonzero correction will be produced.

When the multi-stage method is applied to (5.59), the mth stage becomes

$$Q_{2h}^{(m)} = Q_{2h}^{(0)} - \alpha_m h[R(Q_{2h}^{(m-1)}) + P_{2h}], \tag{5.63}$$

where $R(Q_{2h}^{(m-1)})$ is computed as in (5.36). Note that P_{2h} does not depend on m and remains fixed during the stages. If the present coarse mesh is not the coarsest mesh in the sequence, then one or more iterations of the multi-stage method are performed and the problem is transferred to the next coarser mesh after an additional computation of the residual. The residual and solution are restricted using the operators in (5.57) and (5.58), respectively. When continuing to a coarser mesh, the residual that is restricted must include the forcing term, i.e. $R_{2h}(Q_{2h}) + P_{2h}$.

Once the coarsest grid level is reached, the correction to the solution must be transferred, or *prolonged*, back to the next finer grid. There is an important condition that the transfer operators must satisfy in order to achieve mesh-size independent rates of convergence of the multigrid algorithm, which can be written as [11]:

$$p_{\mathrm{R}} + p_{\mathrm{P}} + 2 > p_{\mathrm{PDE}}, \tag{5.64}$$

where p_{R} and p_{P} are the highest degree polynomials interpolated exactly by the restriction and prolongation operators, respectively, and p_{PDE} is the order of the PDE. For the restriction operator given in (5.57), $p_{\mathrm{R}} = 0$. Therefore, a prolongation based on a piecewise constant interpolation ($p_{\mathrm{P}} = 0$) is adequate for the Euler equations, but

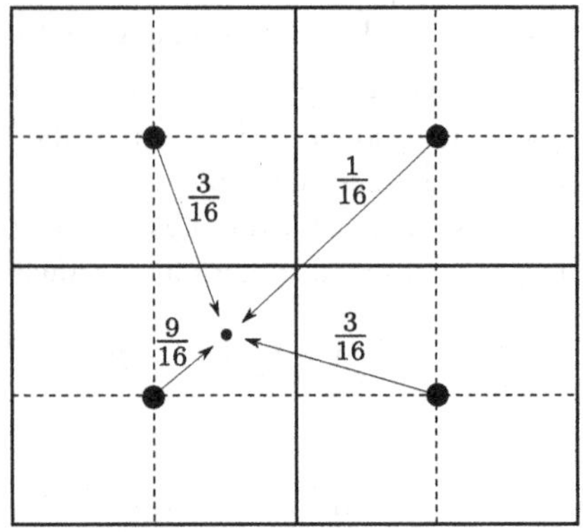

Fig. 5.15 Bilinear prolongation operator for cell-centered scheme in two dimensions

a piecewise linear interpolation ($p_P = 1$) is needed for the Navier-Stokes equations, for which $p_{PDE} = 2$.

The prolongation operation for a cell-centered algorithm in two dimensions is depicted in Fig. 5.15. With bilinear interpolation, the value of the correction ΔQ in each fine mesh cell is calculated based on ΔQ in four coarse mesh cells. The resulting prolongation operator is

$$I_{2h}^{h} \Delta Q = \frac{1}{16}(9\Delta Q_1 + 3\Delta Q_2 + 3\Delta Q_3 + \Delta Q_4), \tag{5.65}$$

where ΔQ_1 is the value in the coarse mesh cell containing the fine mesh cell, ΔQ_2 and ΔQ_3 are the values in the coarse mesh cells that share an edge with the coarse mesh cell containing the fine mesh cell, and ΔQ_4 is the value in the coarse mesh cell that shares only a vertex with the fine mesh cell.

The ΔQ to be prolonged to the fine mesh is the difference between Q_{2h} after completing the iteration or iterations on the coarse mesh and the original $Q_{2h}^{(0)}$ that was restricted to the coarse mesh based on (5.58). Hence we obtain for the corrected Q_h on the fine mesh:

$$Q_h^{(\text{corrected})} = Q_h + I_{2h}^{h}(Q_{2h} - Q_{2h}^{(0)}), \tag{5.66}$$

where Q_h is the value originally computed on the fine mesh (see (5.58)), and I_{2h}^{h} is the prolongation operator given in (5.65).

This basic two-grid framework provides the basis for many variations, known as multigrid cycles, that depend on the number of grids in the sequence and the manner in which they are visited. Figure 5.16 displays two popular cycles, the *V cycle* and the *W cycle*, based on four grids. Downward pointing arrows indicate restriction to a coarser mesh, while upward pointing arrows indicate prolongation to a finer mesh. There are trade-offs between the two cycles, and typically experimentation is needed to determine which is more efficient and robust for a given problem class. In the W cycle, relatively more computations are performed on the coarser grid levels; since

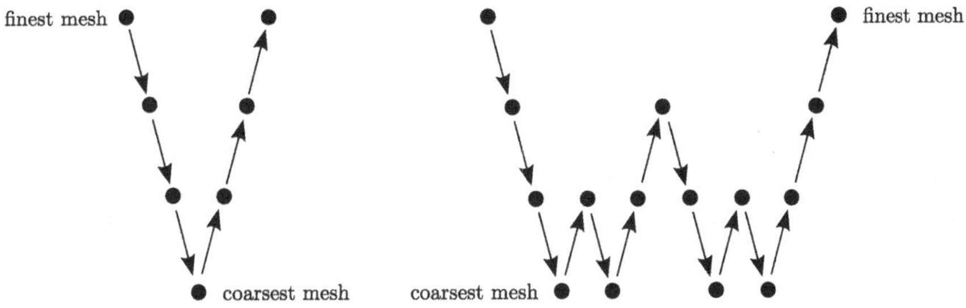

Fig. 5.16 Four-grid V and W multigrid cycles

Fig. 5.17 Full multigrid with four grids

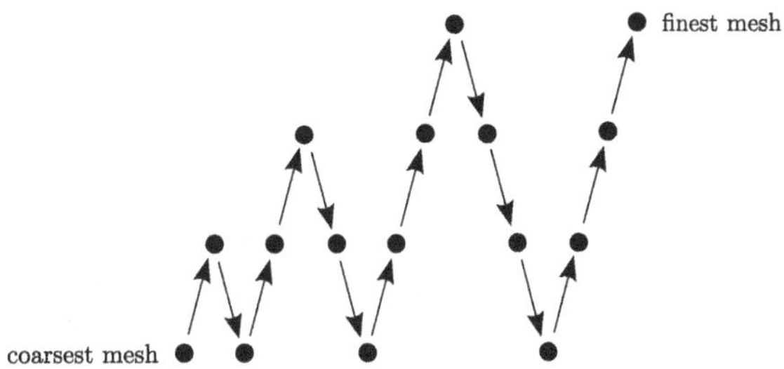

these are inexpensive, the W cycle is often more efficient than the V cycle. Unlike the classical approach to multigrid for linear problems, where the problem is solved exactly on the coarsest mesh, in the present context one simply applies one or more iterations of the multi-stage method on the coarsest mesh. Experiments show that there is typically no benefit to converging further on the coarsest mesh. Similarly, it is rare to see an overall benefit in terms of computational expense in going beyond four or five grids. Within a given cycle, there are also several possible variants. For example, one can apply the multi-stage scheme at each grid level when transferring from the coarse grid levels back to the fine levels, or one can simply add the correction and prolong the result to the next finer grid. Some authors apply an implicit smoothing to the corrections. It is also common to apply various simplifications, such as a lower-order spatial discretization, on the coarser grids. This reduces the computational expense without affecting the converged solution.

Finally, the *full multigrid method* combines the concept of mesh sequencing presented in Sect. 4.5.6 with the multigrid method. Since a sequence of meshes exists as well as a transfer operator from coarse to fine meshes, this is a natural approach. The computation begins on the coarsest mesh in the sequence, on which a number of multi-stage iterations are performed. The solution is transferred to the next finer mesh, and a number of two-grid multigrid cycles are carried out. This solution is transferred to the next finer grid, and a number of three-grid cycles are performed. This process continues until the full cycle is reached, as depicted in Fig. 5.17.

5.4 One-Dimensional Examples

As in Chap. 4, we present examples of the application of the algorithm described in this chapter to the quasi-one-dimensional Euler equations. These examples coincide with the exercises listed at the end of the chapter, giving the reader a benchmark for their results. In the context of a one-dimensional uniform mesh, the implementation of the second-order finite-volume method described in this chapter is very similar to that of the second-order finite-difference method of the previous chapter. Consequently, we will use the same spatial discretization as in Sect. 4.8, but coupled with the explicit multi-stage multigrid algorithm presented in this chapter. Our focus here is on steady flows.

The spatial discretization used to illustrate the performance of the multi-stage multigrid algorithm is node centered. Therefore, the grid transfer operators described in this chapter cannot be used, and we introduce suitable operators for a node-centered scheme in one dimension. The coarse grid is formed by removing every other grid node from the fine mesh. An odd number of nodes should be used to ensure that the boundary nodes are preserved in the coarse mesh. For a sequence of p grids, the finest mesh should have a number of interior nodes equal to some multiple of 2^{p-1} minus one.

The simplest restriction operator is simple injection, where the coarse grid node is assigned the value at the corresponding fine grid node. In a linear weighted restriction, the coarse grid node is assigned a value equal to one-half that at the corresponding fine grid node plus one-quarter of that at each of the neighbours of the fine grid node, which do not exist on the coarse grid. The reader should experiment with these two approaches in order to examine their effect on multigrid convergence. After restricting the solution to the coarse mesh, the boundary values should be reset to satisfy the boundary conditions on the coarse mesh.

For prolongation, linear interpolation gives the following transfer operator. Each fine grid node for which there is a corresponding coarse grid node is assigned the value at that coarse grid node. For fine grid nodes that do not exist on the coarse grid, they receive one-half of the value from each neighbouring coarse grid node. After prolonging the correction to a finer mesh, the boundary values should be reset to satisfy the boundary conditions on the fine mesh.

For the methods presented in this chapter and the previous one, the converged steady solution is independent of the details of the iterative method such as the time step. Since we apply the same spatial discretization as for the results presented in Sect. 4.8, except for a different value of κ_4, the solutions will be nearly identical to those presented previously, as long as the residual is reduced sufficiently. Therefore, we concentrate here only on convergence histories.

The example results presented here are based on the five-stage time-marching method with $\alpha_1 = 1/4$, $\alpha_2 = 1/6$, and $\alpha_3 = 3/8$ and the artificial dissipation computed only on stages 1, 3, and 5. Without residual smoothing, $C_n = 3$. Residual smoothing is applied with $\beta = 0.6$ and $C_n = 7$. The multigrid method is based on the multi-stage method with implicit residual smoothing and the same parameter values.

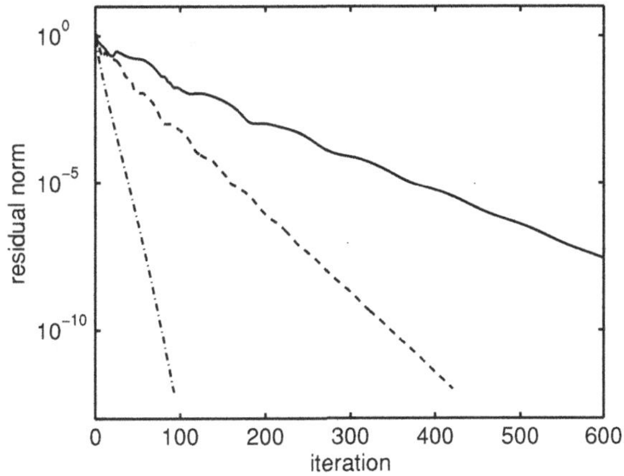

Fig. 5.18 Residual convergence histories for the subsonic channel flow problem with 103 interior nodes using the explicit algorithm with $C_n = 3$ (-), $C_n = 7$ and implicit residual smoothing with $\beta = 0.6$ (- -), and a four-level W multigrid cycle with $C_n = 7$ and implicit residual smoothing with $\beta = 0.6$ (-.)

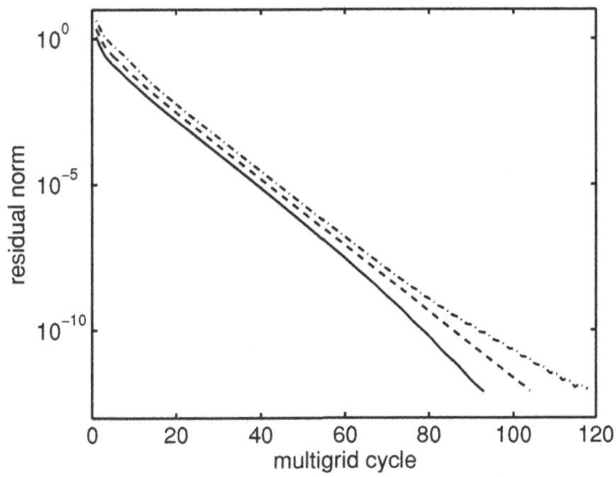

Fig. 5.19 Residual convergence histories for the subsonic channel flow problem with 103 interior nodes (-), 207 interior nodes (- -), and 415 interior nodes (-.) using the explicit algorithm with a W multigrid cycle with $C_n = 7$ and implicit residual smoothing with $\beta = 0.6$. Four grid levels are used on the coarsest mesh, five on the intermediate mesh, and six on the finest mesh

The solution is restricted through simple injection, while linear weighted restriction is used for the residual. For both W and V cycles, the time-marching method is not applied after prolongation except on the finest mesh when the cycle is repeated. The artificial dissipation coefficients are $\kappa_4 = 1/32$ and $\kappa_2 = 0.5$ in all cases.

Figure 5.18 compares the convergence of the explicit algorithm on a single grid without implicit residual smoothing, with implicit residual smoothing, and with a four-level W multigrid cycle for the subsonic channel on a mesh with 103 interior nodes. The norm of the residual of the conservation of mass equation is shown. With the multigrid algorithm, the residual is reduced to below 10^{-12} in 93 multigrid cycles. Figure 5.19 displays the performance of the multigrid algorithm W cycle for varying numbers of grid nodes. Four grid levels are used with 103 interior nodes, five with 207 interior nodes, and six with 413 interior nodes. Thus the coarsest mesh, which has 12 interior nodes, is the same in each case. With this approach, the number of multigrid cycles needed for convergence is nearly independent of the mesh size, as shown in the figure. Figure 5.20 shows that the V cycle does not converge as quickly for this case and requires more cycles as the mesh is refined.

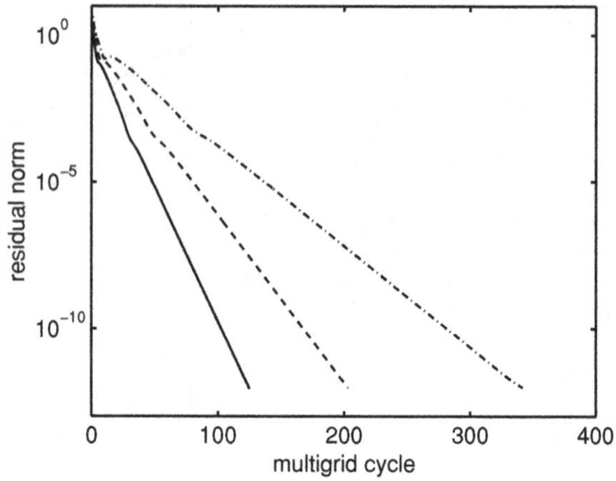

Fig. 5.20 Residual convergence histories for the subsonic channel flow problem with 103 interior nodes (-), 207 interior nodes (- -), and 415 interior nodes (-.) using the explicit algorithm with a V multigrid cycle with $C_{\rm n} = 7$ and implicit residual smoothing with $\beta = 0.6$. Four grid levels are used on the coarsest mesh, five on the intermediate mesh, and six on the finest mesh

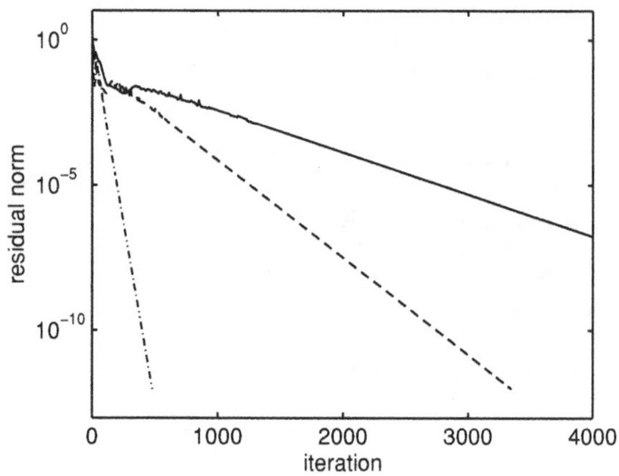

Fig. 5.21 Residual convergence histories for the transonic channel flow problem with 103 interior nodes using the explicit algorithm with $C_{\rm n} = 3$ (-), $C_{\rm n} = 7$ and implicit residual smoothing with $\beta = 0.6$ (- -), and a four-level W multigrid cycle with $C_{\rm n} = 7$ and implicit residual smoothing with $\beta = 0.6$ (-.)

Figures 5.21, 5.22, and 5.23 show the same comparisons for the transonic channel problem. Although the number of iterations or multigrid cycles required for convergence is much higher in this case, the trends are very similar. Implicit residual smoothing improves the convergence rate by a factor close to two. Multigrid is very effective in reducing the number of iterations needed, and the W cycle converges in fewer cycles than the V cycle.

5.5 Summary

The algorithm described in this chapter has the following key features:

- The discretization of the spatial derivatives is accomplished through a second-order cell-centered finite-volume method applied on a structured grid. This approach

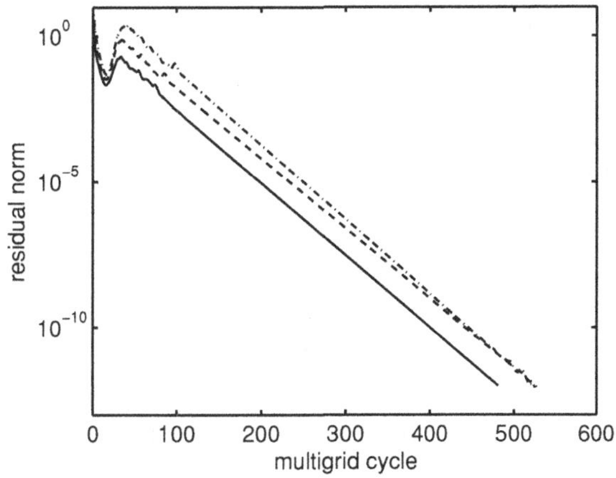

Fig. 5.22 Residual convergence histories for the transonic channel flow problem with 103 interior nodes (-), 207 interior nodes (- -), and 415 interior nodes (-.) using the explicit algorithm with a W multigrid cycle with $C_n = 7$ and implicit residual smoothing with $\beta = 0.6$. Four grid levels are used on the coarsest mesh, five on the intermediate mesh, and six on the finest mesh

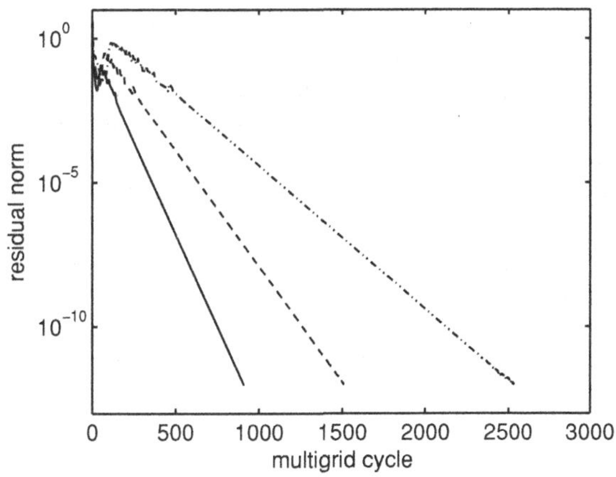

Fig. 5.23 Residual convergence histories for the transonic channel flow problem with 103 interior nodes (-), 207 interior nodes (- -), and 415 interior nodes (-.) using the explicit algorithm with a V multigrid cycle with $C_n = 7$ and implicit residual smoothing with $\beta = 0.6$. Four grid levels are used on the coarsest mesh, five on the intermediate mesh, and six on the finest mesh

can be extended to unstructured grids. Numerical dissipation is added through a nonlinear artificial dissipation scheme that combines a third-order dissipative term in smooth regions of the flow with a first-order term near shock waves. A pressure-based term is used as a shock sensor.

- After discretization in space, the original PDEs are converted to a large system of ODEs. For computations of steady flows, a five-stage explicit method is used in which the artificial dissipation is computed only on stages one, three, and five, and the viscous flux operator is applied only on the first stage. At each stage, the residual is smoothed by application of a scalar tridiagonal implicit operator in each coordinate direction. The multigrid method is applied in order to accelerate convergence to steady state. For computations of unsteady flows, this algorithm can be used within the context of an implicit dual-time-stepping approach.

5.6 Exercises

For related discussion, see Sect. 5.4.

5.1 Write a computer program to apply the explicit multigrid algorithm presented in this chapter to the quasi-one-dimensional Euler equations for the following subsonic problem. $S(x)$ is given by

$$S(x) = \begin{cases} 1 + 1.5\left(1 - \frac{x}{5}\right)^2 & 0 \le x \le 5 \\ 1 + 0.5\left(1 - \frac{x}{5}\right)^2 & 5 \le x \le 10 \end{cases} \tag{5.67}$$

where $S(x)$ and x are in meters. The fluid is air, which is considered to be a perfect gas with $R = 287\ \text{N} \cdot \text{m} \cdot \text{kg}^{-1} \cdot \text{K}^{-1}$, and $\gamma = 1.4$, the total temperature is $T_0 = 300$ K, and the total pressure at the inlet is $p_{01} = 100$ kPa. The flow is subsonic throughout the channel, with $S^* = 0.8$. Use the spatial discretization described in Chap. 4 with the nonlinear scalar artificial dissipation model, since, on a uniform mesh in one dimension, it is essentially the same as that presented in this chapter. Compare your solution with the exact solution computed in Exercise 3.1. Show the convergence history for each case. Experiment with parameters, such as the multigrid cycle (e.g. W and V), the number of grid levels, the Courant number, and the implicit residual smoothing coefficient, to examine their effect on convergence. Find optimal values of the implicit residual smoothing coefficient and the Courant number for rapid and reliable convergence.

5.2 Repeat Exercise 5.1 for a transonic flow in the same channel. The flow is subsonic at the inlet, there is a shock at $x = 7$, and $S^* = 1$. Compare your solution with that calculated in Exercise 3.2.

References

1. Jameson, A., Schmidt, W., Turkel, E.: Numerical solutions of the Euler equations by finite volume methods using Runge-Kutta time-stepping schemes. AIAA Paper 81-1259, 1981
2. Baker, T.J., Jameson, A., Schmidt, W.: A family of fast and robust Euler codes. Princeton University Report MAE 1652, 1984
3. Jameson, A., Baker, T.J.: Multigrid solution of the Euler equations for aircraft configurations. AIAA Paper 84-0093, 1984
4. Jameson, A.: Solution of the Euler equations for two-dimensional transonic flow by a multigrid method. Appl. Math. Comput. **13**(3–4), 327–355 (1983)
5. Jameson, A.: Multigrid algorithms for compressible flow calculations. In: Proceedings of the 2nd European Conference on Multigrid Methods, Lecture Notes in Mathematics 1228, Springer-Verlag, 1986
6. Swanson, R.C., Turkel, E.: On central-difference and upwind schemes. J. Comput. Phys. **101**(2), 292–306 (1992)
7. Swanson, R.C., Turkel, E.: Multistage schemes with multigrid for Euler and Navier-Stokes equations. NASA TP 3631, 1997
8. Warming, R.F., Beam, R.M., Hyett, B.J.: Diagonalization and simultaneous symmetrization of the gas-dynamic matrices. Math. Comput. **29**(132), 1037–1045 (1975)

9. Van der Houwen, P.J.: Construction of Integration Formulas for Initial Value Problems. North Holland, Amsterdam (1977)
10. Lomax, H., Pulliam, T.H., Zingg, D.W.: Fundamentals of Computational Fluid Dynamics. Springer, Berlin (2001)
11. Hemker, P.W.: On the order of prolongations and restrictions in multigrid procedures. J. Comput. Appl. Math. **32**(3), 423–429 (1990)

Chapter 6
Introduction to High-Resolution Upwind Schemes

6.1 Introduction

The algorithms described in the two preceding chapters have been used successfully for the computation of many flow fields. However, in some contexts an additional degree of robustness is required. There are many flow problems, especially those involving complex physics, where positivity and monotonicity preservation is critical. For example, in a hypersonic flow, where the Mach number is much greater than unity, strong shock waves arise. Even a small oscillation in the solution can lead to negative pressures, densities, and temperatures, all of which are unphysical. In the computation of the sound speed for a perfect gas, the square root of these quantities is needed, which is not possible if the quantity is negative. Similarly, oscillations in other regions where the solution is discontinuous or nearly so can also trigger unphysical behavior. The algorithms described thus far were not designed specifically to preserve monotonicity and positivity. For example, the shock sensor (4.83) is based on pressure only, so it cannot sense discontinuities where the pressure is continuous, such as contact surfaces. Moreover, such a sensor is not designed to ensure positivity of a quantity such as turbulent kinetic energy in a turbulence model. High-resolution upwind schemes were developed in order to address the need to maintain positivity and monotonicity of specific quantities consistent with the physics. This chapter provides only a brief introduction to these concepts. The reader is encouraged to complete the exercise at the end of the chapter and to experiment with different strategies. We begin with an introduction to Godunov's method, which is based on a different perspective than the approaches described thus far and has been an influential building block in the development of high-resolution upwind schemes.

T. H. Pulliam and D. W. Zingg, *Fundamental Algorithms in Computational Fluid Dynamics*, Scientific Computation, DOI: 10.1007/978-3-319-05053-9_6,

6.2 Godunov's Method

In our earlier discussion of upwind schemes in Sect. 2.5.2, we introduced flux-vector [1, 2] and flux-difference splitting [3]. Dating back to 1959, Godunov's method [4] provides an alternative approach to upwinding based on the solution of local Riemann problems.

Consider the PDE form of a one-dimensional conservation law:

$$\frac{\partial u}{\partial t} + \frac{\partial f}{\partial x} = 0, \tag{6.1}$$

where $u(x, t)$ is the conserved variable, and $f(u)$ is the flux. The integral form in space on $a \leq x \leq b$ is:

$$\frac{\mathrm{d}}{\mathrm{d}t} \int_a^b u(x, t)\mathrm{d}x = -\{f[u(b, t)] - f[u(a, t)]\}. \tag{6.2}$$

Integrating in time from t_n to t_{n+1} gives

$$\int_a^b u(x, t_{n+1})\mathrm{d}x - \int_a^b u(x, t_n)\mathrm{d}x = -\Delta t\{\bar{f}[u(b, t)] - \bar{f}[u(a, t)]\}, \tag{6.3}$$

where $\bar{f}[u(a, t)]$ is the average flux at $x = a$ over the time interval Δt. Introducing the cell average of the conserved variable, as in Sect. 2.4.1, we have

$$\overline{u_j^n} = \frac{1}{\Delta x} \int_{x_{j-1/2}}^{x_{j+1/2}} u(x, t_n)\mathrm{d}x, \tag{6.4}$$

for a cell where $x_{j-1/2} \leq x \leq x_{j+1/2}$ (see Fig. 2.4). Substituting the cell average into (6.3) with $a = x_{j-1/2}$ and $b = x_{j+1/2}$, we obtain

$$\overline{u_j^{n+1}} - \overline{u_j^n} = -\frac{\Delta t}{\Delta x}\{\bar{f}[u(x_{j+1/2}, t)] - \bar{f}[u(x_{j-1/2}, t)]\}. \tag{6.5}$$

This is an exact statement of the conservation law showing that the change in the cell average of the conserved variable over the time interval Δt is dependent upon the average flux at the cell boundaries over the time interval.

The first step in Godunov's method is to reconstruct the solution in each cell based on the cell-averaged solution. A piecewise-constant reconstruction is used, as in Sect. 2.4.2, given by

$$u(x, t_n) = \overline{u_j^n} \quad x_{j-1/2} \leq x \leq x_{j+1/2}. \tag{6.6}$$

The cell boundaries lie at $x_{j-1/2}$ and $x_{j+1/2}$. As in Sect. 2.4.2, there are two states at each boundary, a left state and a right state, as a result of the reconstruction in each cell. For example, at $x_{j+1/2}$, we have

$$u^{\mathrm{L}}_{j+1/2} = \bar{u}_j \quad \text{and} \quad u^{\mathrm{R}}_{j+1/2} = \bar{u}_{j+1}. \tag{6.7}$$

In Sect. 2.4.2, we resolved this discontinuity by taking the average of the fluxes on either side of the boundary, leading to a nondissipative finite-volume scheme analogous to a centered difference scheme.

In Godunov's method, the discontinuity in the solution is resolved through the exact solution of the local Riemann problem. We define

$$u^*\left(\frac{x}{t}, u^{\mathrm{L}}, u^{\mathrm{R}}\right) \tag{6.8}$$

as the exact solution of the Riemann problem with initial condition

$$\begin{aligned} u &= u^{\mathrm{L}} \quad x < 0 \\ u &= u^{\mathrm{R}} \quad x \geq 0. \end{aligned} \tag{6.9}$$

In the case of the Euler equations, this represents the solution to the one-dimensional flow problem arising when a diaphragm separating two distinct fluid states is removed at $t = 0$, i.e. a generalization of the shock-tube problem discussed in Sect. 3.3.2. The solution is characterized by the fact that it is constant along rays emanating from the origin of the $x - t$ plane and consequently depends on x/t, rather than x and t independently.

With the piecewise-constant reconstruction, the Riemann problem originating at time level n and $x_{j+1/2}$, i.e. the right boundary of the cell, has the left state $\overline{u^n_j}$ and the right state $\overline{u^n_{j+1}}$. Hence the Riemann problem can be written as

$$u(x,t) = u^*\left(\frac{x - x_{j+1/2}}{t - t_n}, \overline{u^n_j}, \overline{u^n_{j+1}}\right). \tag{6.10}$$

Similarly, the Riemann problem originating at the left boundary of the cell is given by

$$u(x,t) = u^*\left(\frac{x - x_{j-1/2}}{t - t_n}, \overline{u^n_{j-1}}, \overline{u^n_j}\right). \tag{6.11}$$

In general, each Riemann problem will include both left and right moving waves. The solution in cell j is determined by the left moving waves from the Riemann problem with its origin at $x_{j+1/2}$ and the right moving waves from the Riemann problem with its origin at $x_{j-1/2}$. These solutions are no longer valid once the Riemann problems interact, which occurs when the fastest left moving wave from the Riemann problem with origin at $x_{j+1/2}$ meets the fastest right moving wave from the Riemann problem with origin at $x_{j-1/2}$. To ensure that the Riemann problems do not interact, we limit the time step Δt according to

$$|a_{\max}|\Delta t < \frac{\Delta x}{2}, \tag{6.12}$$

where $a_{\max}$ is the highest wave speed in the system.

With this limit on Δt, the solution at t_{n+1} is determined by the Riemann problem originating from $x_{j-1/2}$ for $x_{j-1/2} \le x \le x_j$. Similarly, the solution for $x_j \le x \le x_{j+1/2}$ is determined by the Riemann problem originating from $x_{j+1/2}$. We are now in a position to determine the cell average state variable in cell j at t_{n+1} as follows:

$$\overline{u_j^{n+1}} = \frac{1}{\Delta x}\int_{x_{j-1/2}}^{x_{j+1/2}} u(x, t_{n+1})\mathrm{d}x = \frac{1}{\Delta x}\left[\int_{x_{j-1/2}}^{x_j} u^*\left(\frac{x - x_{j-1/2}}{\Delta t}, \overline{u_{j-1}^n}, \overline{u_j^n}\right)\mathrm{d}x \right.$$
$$\left. + \int_{x_j}^{x_{j+1/2}} u^*\left(\frac{x - x_{j+1/2}}{\Delta t}, \overline{u_j^n}, \overline{u_{j+1}^n}\right)\mathrm{d}x\right], \tag{6.13}$$

where the first term on the right-hand side is associated with the Riemann problem at the left boundary of the cell and the second term with the Riemann problem at the right boundary of the cell.

As a simple example to clarify these concepts, consider the linear convection equation:

$$\frac{\partial u}{\partial t} + a\frac{\partial u}{\partial x} = 0 \tag{6.14}$$

with $a > 0$. With a known solution at $t = t_n$ given by $u_n(x)$, in the absence of boundaries the exact solution is

$$u(x, t) = u_n\left(x - a(t - t_n)\right), \tag{6.15}$$

which represents the waveform at t_n propagating to the right with speed a. This exact solution provides the solution to the Riemann problem. In this case there are no left moving waves, so the time step must satisfy

$$a\Delta t \le \Delta x \tag{6.16}$$

in order to prevent interacting Riemann problems.

With a piecewise-constant reconstruction, the solution in cell $j - 1$ at t_n is given by

$$u(x, t_n) = \overline{u_{j-1}^n}, \tag{6.17}$$

and that in cell j is

$$u(x, t_n) = \overline{u_j^n}. \tag{6.18}$$

After a time step Δt, the solution from cell $j-1$ has convected into cell j a distance $a\Delta t$, so the value of u in cell j at t_{n+1} from $x_{j-1/2}$ to $x_{j-1/2}+a\Delta t$ is $\overline{u_{j-1}^n}$. The solution in the remainder of cell j at t_{n+1}, from $x_{j-1/2}+a\Delta t$ to $x_{j+1/2}$ is $\overline{u_j^n}$. By analogy to (6.13), the cell average in cell j can thus be updated according to

$$\begin{aligned}\overline{u_j^{n+1}} &= \frac{1}{\Delta x}\int_{x_{j-1/2}}^{x_{j+1/2}} u(x,t_{n+1})\mathrm{d}x \\ &= \frac{1}{\Delta x}\left[a\Delta t\overline{u_{j-1}^n} + (\Delta x - a\Delta t)\overline{u_j^n}\right] \\ &= \overline{u_j^n} - \frac{a\Delta t}{\Delta x}(\overline{u_j^n} - \overline{u_{j-1}^n}).\end{aligned} \tag{6.19}$$

This we can recognize immediately as first-order backward differencing in space combined with the explicit Euler time-marching method, and the time step restriction (6.16) is consistent with the stability bound for this scheme.

This example sheds some light on Godunov's method. Despite the use of an exact solution, the resulting numerical method is first order in both time and space. The first-order explicit Euler method arises naturally from the determination of the solution at t_{n+1} explicitly from the solution at t_n. First-order accuracy in space is a consequence of the piecewise-constant reconstruction. The key point is that Godunov's method has produced an upwind scheme. If we repeat the exercise with $a < 0$, then a forward difference scheme results. Hence the use of an exact Riemann solution by Godunov's method naturally produces an upwind scheme; this result applies to systems of equations as well.

Godunov's method can be simplified by exploiting the fact that the solution to a Riemann problem is constant along rays emanating from the origin of the problem in the $x-t$ plane. As a result, the Riemann problem originating at $x = x_{j-1/2}, t = t_n$ does not vary with time at $x = x_{j-1/2}$. Similarly, the Riemann problem originating at $x = x_{j+1/2}, t = t_n$ does not vary with time at $x = x_{j+1/2}$. Recall from (6.5) that the change in the cell average of the conserved variable over the time interval Δt is determined by the average flux at the cell boundaries over the time interval. Hence we obtain

$$\overline{u_j^{n+1}} = \overline{u_j^n} - \frac{\Delta t}{\Delta x}\left[f(u^*(0,\overline{u_j^n},\overline{u_{j+1}^n})) - f(u^*(0,\overline{u_{j-1}^n},\overline{u_j^n}))\right], \tag{6.20}$$

where the flux function is evaluated at the state determined from the Riemann solution at the value of x corresponding to the origin of the Riemann problem, and the average in time is trivial because the state is constant over the time interval at each cell boundary, i.e. at $x_{j-1/2}$ and $x_{j+1/2}$.

Based on the above, we define the numerical flux function $\hat{f}$ for Godunov's method as

$$\hat{f}_{j+1/2} = f(u^*(0, \overline{u_j^n}, \overline{u_{j+1}^n})). \tag{6.21}$$

The definition of $\hat{f}_{j-1/2}$ is found simply by decrementing the spatial index by one,[1] i.e.

$$\hat{f}_{j-1/2} = f(u^*(0, \overline{u_{j-1}^n}, \overline{u_j^n})). \tag{6.22}$$

With this definition of the Godunov numerical flux function we can write Godunov's method in the generic finite-volume form:

$$\overline{u_j^{n+1}} = \overline{u_j^n} - \frac{\Delta t}{\Delta x}(\hat{f}_{j+1/2} - \hat{f}_{j-1/2}). \tag{6.23}$$

Moreover, returning to (6.2), we can write the generic semi-discrete form:

$$\frac{\mathrm{d}\overline{u_j}}{\mathrm{d}t} = -\frac{1}{\Delta x}(\hat{f}_{j+1/2} - \hat{f}_{j-1/2}). \tag{6.24}$$

For the example of the linear convection equation, but with a of arbitrary sign, the Godunov numerical flux function is

$$\hat{f}_{j+1/2} = \frac{1}{2}(a + |a|)\overline{u_j^n} + \frac{1}{2}(a - |a|)\overline{u_{j+1}^n}. \tag{6.25}$$

It is left as an exercise for the reader to show that this is entirely equivalent to (2.89) and (2.90). As a nonlinear example, consider the Burgers equation:

$$\frac{\partial u}{\partial t} + \frac{1}{2}\frac{\partial u^2}{\partial x} = 0. \tag{6.26}$$

In this case one can show that Godunov's flux function is given by [5][2]:

$$\hat{f}_{j+1/2} = \begin{cases} \frac{1}{2}u_{j+1}^2 & \text{if} \quad u_j, u_{j+1} \text{ are both } \leq 0 \\ \frac{1}{2}u_j^2 & \text{if} \quad u_j, u_{j+1} \text{ are both } \geq 0 \\ 0 & \text{if} \quad u_j \leq 0 \leq u_{j+1} \\ \frac{1}{2}u_j^2 & \text{if} \quad u_j > 0 \geq u_{j+1} \text{ and } |u_j| \geq |u_{j+1}| \\ \frac{1}{2}u_{j+1}^2 & \text{if} \quad u_j \geq 0 > u_{j+1} \text{ and } |u_j| \leq |u_{j+1}| \end{cases}. \tag{6.27}$$

This numerical flux function is based on the exact Riemann solution to the scalar conservation law and provides a useful reference for assessing approximate solutions to the Riemann problem such as the one introduced in the next section.

---

[1] This is a necessary property of a conservative scheme.

[2] Henceforth we drop the bars denoting cell average quantities for convenience.

6.3 Roe's Approximate Riemann Solver

In the previous section we saw that, despite the use of an exact Riemann solution, Godunov's method is first-order accurate in space as a result of the piecewise-constant reconstruction. This motivates the idea of reducing the computational expense by using an approximate Riemann solver to provide the desired upwinding. There are several such approximate Riemann solvers for the Euler equations; the approach of Roe [3] is particularly widely used.

In Roe's approximate Riemann solver, the conservation law is locally linearized. For example, consider the quasi-linear form of the Euler equations (3.26):

$$\frac{\partial Q}{\partial t} + A\frac{\partial Q}{\partial x} = 0, \tag{6.28}$$

where the flux Jacobian A is a function of Q. If we linearize about some state $\bar{Q}$ and define $\bar{A} = A(\bar{Q})$, we can write the locally linearized form of the Euler equations as

$$\frac{\partial Q}{\partial t} + \bar{A}\frac{\partial Q}{\partial x} = 0. \tag{6.29}$$

Since $\bar{A}$ is independent of Q, the equations can be decoupled into three equations in the form of the linear convection equation, as described in Sect. 4.6.1. The exact solution to this linearized problem is readily obtained in terms of the eigensystem of $\bar{A}$.

Recall that for a Riemann problem, there are initially two states. Roe chose the average state for the linearization such that it satisfies

$$f^R - f^L = A(\bar{Q})(Q^R - Q^L), \tag{6.30}$$

where $f = AQ$ is the flux. This ensures satisfaction of the Rankine-Hugoniot relations at shock waves and full upwinding in supersonic flows, where all of the eigenvalues of the flux Jacobian have the same sign. For the Euler equations, the state that satisfies this relation, known as the *Roe-average state*, is given by [3]:

$$\begin{aligned}
\bar{\rho} &= \sqrt{\rho^L \rho^R} \\
\bar{u} &= \frac{(u\sqrt{\rho})^L + (u\sqrt{\rho})^R}{\sqrt{\rho^L} + \sqrt{\rho^R}} \\
\bar{H} &= \frac{(H\sqrt{\rho})^L + (H\sqrt{\rho})^R}{\sqrt{\rho^L} + \sqrt{\rho^R}},
\end{aligned} \tag{6.31}$$

where H is the total enthalpy.

Next we will determine the numerical flux function for the Roe scheme with a piecewise-constant reconstruction in each cell, which gives $Q^L_{j+1/2} = Q_j$ and

$Q^R_{j+1/2} = Q_{j+1}$. As a result of the local linearization of the one-dimensional Euler equations, they are decoupled into three equations in the form of the linear convection equation as follows (see Sect. 4.6.1):

$$\frac{\partial W}{\partial t} + \Lambda \frac{\partial W}{\partial x} = 0, \tag{6.32}$$

where $W = X^{-1}Q$ are the characteristic variables, X^{-1} has the left eigenvectors of $\bar{A}$ as its rows, and Λ is a diagonal matrix containing the eigenvalues of $\bar{A}$, which is based on the Roe-average state determined from the cell averages Q_j and Q_{j+1} on either side of the cell interface at $x_{j+1/2}$. Applying (6.25) to each equation individually and then recoupling by premultiplying by X gives

$$\begin{aligned}
\hat{f}_{j+1/2} &= X\left[\frac{1}{2}(\Lambda + |\Lambda|)W_j + \frac{1}{2}(\Lambda - |\Lambda|)W_{j+1}\right] \\
&= X\left[\frac{1}{2}(\Lambda + |\Lambda|)X^{-1}Q_j + \frac{1}{2}(\Lambda - |\Lambda|)X^{-1}Q_{j+1}\right] \\
&= \frac{1}{2}X\Lambda X^{-1}(Q_j + Q_{j+1}) + \frac{1}{2}X|\Lambda|X^{-1}(Q_j - Q_{j+1}) \\
&= \frac{1}{2}\bar{A}(Q_j + Q_{j+1}) + \frac{1}{2}|\bar{A}|(Q_j - Q_{j+1}) \\
&= \frac{1}{2}(f_j + f_{j+1}) + \frac{1}{2}|\bar{A}|(Q_j - Q_{j+1}),
\end{aligned} \tag{6.33}$$

where X is the matrix of right eigenvectors of $\bar{A}$, and $|A| = X|\Lambda|X^{-1}$, consistent with Sect. 2.5.2. The last step depends on the property of the Roe average given in (6.30) as well as the fact that the Euler equations are homogeneous of order one [6] and hence $f(-Q) = -f(Q)$. Note the similarity to the flux-difference-split scheme given in (2.101), which was derived for the linear, constant-coefficient case.

In the scalar case, the Roe numerical flux function becomes

$$\hat{f}_{j+1/2} = \frac{1}{2}(f_j + f_{j+1}) - \frac{1}{2}|\bar{a}_{j+1/2}|(u_{j+1} - u_j), \tag{6.34}$$

where

$$\bar{a}_{j+1/2} = \begin{cases} \frac{f_{j+1}-f_j}{u_{j+1}-u_j} & \text{if } u_{j+1} \neq u_j \\ a(u_j) & \text{if } u_{j+1} = u_j \end{cases}, \tag{6.35}$$

which is the Rankine-Hugoniot relation for the speed of propagation of a discontinuity. For the Burgers equation ($f(u) = \frac{1}{2}u^2$) one obtains

$$\bar{a}_{j+1/2} = \frac{\frac{u_{j+1}^2}{2} - \frac{u_j^2}{2}}{u_{j+1} - u_j} = \frac{1}{2}(u_j + u_{j+1}). \tag{6.36}$$

Substituting (6.36) into (6.34) gives the flux function for the Roe scheme applied to the Burgers equation:

$$\hat{f}_{j+1/2} = \frac{1}{2}(\frac{1}{2}u_j^2 + \frac{1}{2}u_{j+1}^2) - \frac{1}{2}|\frac{1}{2}(u_j + u_{j+1})|(u_{j+1} - u_j) \tag{6.37}$$

$$= \begin{cases} \frac{1}{2}u_{j+1}^2 & \text{if } \bar{a}_{j+1/2} \leq 0 \\ \frac{1}{2}u_j^2 & \text{if } \bar{a}_{j+1/2} > 0 \end{cases}. \tag{6.38}$$

This differs from the numerical flux function of the Godunov scheme (6.27) only when $u_j < 0 < u_{j+1}$; Godunov's flux function is equal to zero for this case. As a result, the Roe scheme permits expansion shocks, and a simple *entropy fix* is commonly used to overcome this. For example, for the Euler equations, such a fix can be obtained by replacing the eigenvalues $\lambda = u + a$ and $\lambda = u - a$ with

$$\frac{1}{2}\left(\frac{\lambda^2}{\epsilon} + \epsilon\right), \tag{6.39}$$

if they are less than or equal to ϵ, where ϵ is a small parameter [7]. For values of λ greater than ϵ, the values are not altered. This prevents these eigenvalues from being zero at sonic points, where $|u| = a$, thereby preventing the development of expansion shocks.

6.4 Higher-Order Reconstruction

The schemes discussed thus far in this chapter have been based on a piecewise-constant reconstruction, which restricts their accuracy to first-order in space. The numerical flux functions associated with the Godunov and Roe schemes with piecewise-constant reconstruction can be written in the form

$$\hat{f}_{j+1/2} = \hat{f}(u_j, u_{j+1}). \tag{6.40}$$

In order to enable second- and higher-order accuracy, we can generalize this to

$$\hat{f}_{j+1/2} = \hat{f}(u_{j+1/2}^L, u_{j+1/2}^R), \tag{6.41}$$

where $u_{j+1/2}^L$ and $u_{j+1/2}^R$ are the left and right states at $x_{j+1/2}$ determined from the reconstruction. Together with the semi-discrete form (6.24), this enables higher-order accuracy in both space and time.

There exist many different approaches to reconstruction; we will describe a widely used approach here [5, 8]. Given cell average data, reconstruction provides an approximation to the behavior of the function within each cell. Piecewise-constant reconstruction is the simplest form, and, as we have seen, restricts the resulting upwind scheme to first-order spatial accuracy. With a centered flux function one can obtain a second-order scheme from a piecewise constant reconstruction, but the scheme is nondissipative. With each increase in the degree of the polynomial used in the reconstruction, the order of the upwind finite-volume scheme also increases. Hence second-order accuracy can be obtained from a piecewise-linear reconstruction, third-order from a piecewise-quadratic reconstruction, and so on. Here we describe one-dimensional piecewise reconstructions up to quadratic that use data equivalently from neighboring cells on either side of the cell in question.

In a piecewise-constant reconstruction, there is only one degree of freedom, and that is determined by the cell average quantity. For a piecewise-linear reconstruction, there is one additional degree of freedom, the slope. It is determined through a second-order centered difference using the cell average data of the two neighboring cells. For a piecewise quadratic reconstruction, there are three degrees of freedom. These are determined by requiring the quadratic function to have the given average in the cell in question as well as the two neighboring cells. This approach was presented in Sect. 2.4.2 (see (2.73), (2.74), and (2.75)).

The following function includes these three reconstructions valid for cell j, i.e. for $x_{j-1/2} \leq x \leq x_{j+1/2}$:

$$\begin{aligned} u(x) = \overline{u_j} &+ \alpha \left(\frac{\overline{u_{j+1}} - \overline{u_{j-1}}}{2\Delta x} \right)(x - x_j) \\ &+ \beta \left(\frac{\overline{u_{j+1}} - 2\overline{u_j} + \overline{u_{j-1}}}{2\Delta x^2} \right) [(x - x_j)^2 - \frac{\Delta x^2}{12}], \end{aligned} \tag{6.42}$$

where the overbars have been included to emphasize that these are cell averages and will be dropped for the remainder of this section. The piecewise-constant reconstruction is obtained with $\alpha = \beta = 0$, piecewise-linear with $\alpha = 1$, $\beta = 0$, and piecewise-quadratic with $\alpha = \beta = 1$. β are sometimes used with $\alpha = 1$; these produce second-order upwind schemes.

In order to determine a numerical flux at $x_{j+1/2}$ from (6.41), we require $u^L_{j+1/2}$ and $u^R_{j+1/2}$. To find the left state at $x_{j+1/2}$, one must substitute $x = x_j + \Delta x/2$ into (6.42). This gives

$$u^L_{j+1/2} = u_j + \frac{1}{4}[(\alpha - \beta/3)(u_j - u_{j-1}) + (\alpha + \beta/3)(u_{j+1} - u_j)]. \tag{6.43}$$

The right state at $x_{j+1/2}$ is determined from the piecewise reconstruction in cell $j + 1$ by incrementing the indices appearing in (6.42) by one and substituting $x = x_{j+1} - \Delta x/2$. We obtain

$$u^R_{j+1/2} = u_{j+1} + \frac{1}{4}[(\alpha + \beta/3)(u_{j+1} - u_j) + (\alpha - \beta/3)(u_{j+2} - u_{j+1})]. \tag{6.44}$$

The left and right states given by (6.43) and (6.44) can be substituted into (6.41) and finally into the semi-discrete form (6.24) to which a time-marching method can be applied to advance the solution in time.

In order to further understand these reconstructions, consider the linear convection equation with positive a and an upwind flux function, $\hat{f}_{j+1/2} = au^L_{j+1/2}$. We have already seen in (6.19) that the piecewise-constant reconstruction leads to first-order backward differencing in space in this case. With the piecewise-linear reconstruction ($\alpha = 1$, $\beta = 0$), we obtain

$$u^L_{j+1/2} = u_j + \frac{1}{4}(u_{j+1} - u_{j-1}). \tag{6.45}$$

This leads to the following semi-discrete form:

$$\left(\frac{\mathrm{d}u}{\mathrm{d}t}\right)_j = -\frac{a}{4\Delta x}(u_{j+1} + 3u_j - 5u_{j-1} + u_{j-2}). \tag{6.46}$$

This spatial operator is second-order accurate. Finally, with the piecewise-quadratic reconstruction ($\alpha = \beta = 1$), the left state at $x_{j+1/2}$ is

$$u^L_{j+1/2} = \frac{1}{6}(2u_{j+1} + 5u_j - u_{j-1}), \tag{6.47}$$

and thus

$$\left(\frac{\mathrm{d}u}{\mathrm{d}t}\right)_j = -\frac{a}{6\Delta x}(2u_{j+1} + 3u_j - 6u_{j-1} + u_{j-2}). \tag{6.48}$$

This is a third-order upwind-biased operator in space [see 2.18].

6.5 Conservation Laws and Total Variation

Consider a scalar conservation law in one dimension:

$$\frac{\partial u}{\partial t} + \frac{\partial f(u)}{\partial x} = 0. \tag{6.49}$$

The exact solution is constant along characteristic lines given by

$$\frac{\mathrm{d}x}{\mathrm{d}t} = a(u) = \frac{\partial f}{\partial u} \tag{6.50}$$

unless the characteristics intersect to form a shock wave. For an initial value problem, i.e. no influence from boundaries, the total variation of a differentiable solution between any pairs of characteristics is conserved [9], where the total variation is defined as

$$TV(u(x,t)) = \int_{-\infty}^{\infty} \left| \frac{\partial u}{\partial x} \right| \mathrm{d}x. \tag{6.51}$$

In the presence of discontinuities, the total variation is nonincreasing in time, i.e.

$$TV(u(x,t_0+t)) \leq TV(u(x,t_0)), \tag{6.52}$$

if the discontinuities satisfy an entropy inequality [9], where t_0 is the initial time. As a consequence of this, *local maxima do not increase, local minima do not decrease, and monotonic solutions remain monotonic, i.e. no new extrema are created.*

Designing numerical schemes such that the numerical solution retains these properties of the exact solution brings the following benefits:

- Robustness: a nonincreasing total variation precludes the generation of spurious oscillations and ensures that quantities such as density and pressure that are initially positive will remain so.
- Stability: if local minima cannot decrease and local maxima cannot increase, then the solution will remain bounded.

6.6 Monotone and Monotonicity-Preserving Schemes

Consider a conservative discretization of the conservation law written as

$$\begin{aligned} u_j^{n+1} &= u_j^n - \frac{\Delta t}{\Delta x}(\hat{f}_{j+1/2} - \hat{f}_{j-1/2}) \\ &= H(u_{j-l}^n, u_{j-l+1}^n, \ldots, u_{j+l}^n), \end{aligned} \tag{6.53}$$

where

$$\hat{f}_{j+1/2} = \hat{f}(u_{j-l+1}, \ldots, u_{j+l}), \tag{6.54}$$

and l depends on the scheme. The discrete method is *monotone* if H is a monotone increasing function of each of its arguments [10], i.e.

$$\frac{\partial H}{\partial u_i}(u_{-l}, \ldots, u_{+l}) \geq 0 \quad \text{for all} \quad -l \leq i \leq l. \tag{6.55}$$

This is a strong condition that ensures that the numerical solution has the monotonicity properties described above. However, such schemes are limited to first-order accuracy.

As a first example, consider the scheme for the numerical solution of the linear convection equation with positive a resulting from the combination of first-order backward differencing in space and the explicit Euler time-marching method, which is given by

$$u_j^{n+1} = C_{\text{n}} u_{j-1}^n + (1 - C_{\text{n}}) u_j^n, \tag{6.56}$$

where

$$C_{\text{n}} = \frac{a \Delta t}{\Delta x} \tag{6.57}$$

is the Courant number. The scheme is stable for $0 < C_{\text{n}} \leq 1$. Applying the condition (6.55) gives

$$\frac{\partial H}{\partial u_{j-1}} = C_{\text{n}}, \quad \frac{\partial H}{\partial u_j} = 1 - C_{\text{n}}. \tag{6.58}$$

These are both nonnegative for Courant numbers in the stable range; hence the scheme is monotone in this range.

Next consider the Lax-Wendroff method given by [11]:

$$\begin{aligned} u_j^{n+1} &= u_j^n - \frac{1}{2} \frac{ah}{\Delta x} (u_{j+1}^n - u_{j-1}^n) \\ &+ \frac{1}{2} \left(\frac{ah}{\Delta x} \right)^2 (u_{j+1}^n - 2u_j^n + u_{j-1}^n). \end{aligned} \tag{6.59}$$

It can be written in the form of (6.53) as follows

$$u_j^{n+1} = \frac{C_{\text{n}}}{2}(1 + C_{\text{n}}) u_{j-1}^n + (1 - C_{\text{n}}^2) u_j^n + \frac{C_{\text{n}}}{2}(C_{\text{n}} - 1) u_{j+1}^n. \tag{6.60}$$

This method is also stable for $0 < C_{\text{n}} \leq 1$. Although the coefficients of u_{j-1}^n and u_j^n are nonnegative in this range of Courant numbers, the coefficient of u_{j+1}^n is negative unless $C_{\text{n}} = 1$. Therefore the Lax-Wendroff method is not monotone. This is of course unsurprising given that the Lax-Wendroff method is second order in time and space.

The restriction of monotone schemes to first-order accuracy is excessively limiting, and a weaker condition is needed. To this end, Harten [12] defined the *monotonicity preserving* property. A scheme is monotonicity preserving if monotonicity of u^n guarantees monotonicity of u^{n+1}, where a solution u is monotonic if

$$\min(u_{j-1}, u_{j+1}) \leq u_j \leq \max(u_{j-1}, u_{j+1}) \quad \text{for all } j. \tag{6.61}$$

A monotonicity-preserving scheme is sufficient to ensure that the numerical solution has the following properties:

- No new extrema are created.
- Local maxima are nonincreasing.
- Local minima are nondecreasing.

All monotone schemes are monotonicity preserving, but the converse is not true.

All *linear* monotonicity-preserving schemes are at most first-order accurate [4, 12]. Consequently, to be of order higher than first, a monotonicity-preserving scheme must be *nonlinear*. In a nonlinear scheme the coefficients of the scheme are dependent upon the solution. For example, the schemes described in the preceding chapters involving the use of a pressure sensor are nonlinear schemes, because the amount of first-order numerical dissipation added is dependent on the pressure.

6.7 Total-Variation-Diminishing Conditions

In Sect. 6.5, we introduced the idea that the total variation is nonincreasing in the exact solution to a scalar conservation law, where the total variation is defined in (6.51). This suggests that a suitable design condition for a numerical scheme to be monotonicity preserving is to require that it be total variation diminishing (or TVD). For this purpose we define a discrete total variation as

$$TV_{\mathrm{d}}(u) = \sum_{-\infty}^{\infty} |u_j - u_{j-1}|. \tag{6.62}$$

All monotone schemes are TVD, and all TVD schemes are monotonicity preserving [12]. Therefore, TVD schemes provide the properties we seek and are subject to the constraint that to be of order higher than first, they must be nonlinear.

To write the TVD conditions, consider a conservative scheme in the following semi-discrete form

$$\frac{\mathrm{d}u_j}{\mathrm{d}t} = -\frac{1}{\Delta x}(\hat{f}_{j+1/2} - \hat{f}_{j-1/2}). \tag{6.63}$$

This can be rewritten in the form

$$\frac{\mathrm{d}u_j}{\mathrm{d}t} = \frac{1}{\Delta x}[C^-_{j+1/2}(u_{j+1} - u_j) - C^+_{j-1/2}(u_j - u_{j-1})]. \tag{6.64}$$

It is important to recognize that any dependence on $u_{j\pm 2}$, $u_{j\pm 3}$, etc., is contained in the $C^{\pm}$ coefficients—this will become clearer in our second example below.

The TVD conditions are [12]:

$$C^{-}_{j+1/2} \geq 0 \quad \text{and} \quad C^{+}_{j-1/2} \geq 0. \tag{6.65}$$

When the semi-discrete form is advanced in time with the explicit Euler time-marching method, one obtains

$$u^{n+1}_j = u^n_j + \frac{\Delta t}{\Delta x}[C^{-}_{j+1/2}(u^n_{j+1} - u^n_j) - C^{+}_{j-1/2}(u^n_j - u^n_{j-1})]. \tag{6.66}$$

For this fully-discrete form, an additional TVD condition is introduced [12]:

$$1 - \frac{\Delta t}{\Delta x}(C^{-}_{j+1/2} + C^{+}_{j-1/2}) \geq 0. \tag{6.67}$$

Further conditions exist if higher than first-order accuracy in time is sought. For example, strong-stability-preserving schemes have been developed for this purpose [13].

As an example, consider the combination of first-order backward differencing in space and the explicit Euler time-marching method applied to the linear convection equation with positive a. We showed previously that this scheme is monotone for Courant numbers in the stable range $0 < C_{\text{n}} \leq 1$. Writing this scheme in the form of (6.66), we obtain

$$\begin{aligned} u^{n+1}_j &= u^n_j - \frac{a\Delta t}{\Delta x}(u^n_j - u^n_{j-1}) \\ &= u^n_j + \frac{\Delta t}{\Delta x}[0 \cdot (u_{j+1} - u_j) - a(u_j - u_{j-1})], \end{aligned} \tag{6.68}$$

giving $C^{-}_{j+1/2} = 0$ and $C^{+}_{j-1/2} = a > 0$. Hence the TVD conditions (6.65) are satisfied, and (6.67) is satisfied for Courant numbers in the stable range, so the scheme is TVD. This is consistent with the fact that all monotone schemes are TVD.

As a second example, consider the semi-discrete form arising from second-order backward differencing the linear convection equation with positive a:

$$\begin{aligned} \frac{\mathrm{d}u_j}{\mathrm{d}t} &= -\frac{a}{2\Delta x}(3u_j - 4u_{j-1} + u_{j-2}) \\ &= \frac{1}{\Delta x}\left[0 \cdot (u_{j+1} - u_j) - \frac{a}{2}(3(u_j - u_{j-1}) - (u_{j-1} - u_{j-2}))\right] \\ &= \frac{1}{\Delta x}\left[0 \cdot (u_{j+1} - u_j) - \frac{a}{2}\left(3 - \frac{u_{j-1} - u_{j-2}}{u_j - u_{j-1}}\right)(u_j - u_{j-1})\right]. \end{aligned} \tag{6.69}$$

Hence

$$C^{-}_{j+1/2} = 0, \quad C^{+}_{j-1/2} = \frac{a}{2}\left(3 - \frac{u_{j-1} - u_{j-2}}{u_j - u_{j-1}}\right). \tag{6.70}$$

The second TVD condition in (6.65) is violated when

$$\frac{u_{j-1} - u_{j-2}}{u_j - u_{j-1}} > 3, \tag{6.71}$$

showing that this scheme is not TVD in general. Again this is not unexpected, as this is a linear second-order scheme, and such schemes cannot be TVD. In the next section, we will see that ratios of the form (6.71) are important in the development of TVD schemes.

The second-order backwards difference operator gives an approximation to $\partial u/\partial x$ at node j based on the slope of the parabola passing through the three points (x_{j-2}, u_{j-2}), (x_{j-1}, u_{j-1}), (x_j, u_j). If $u_j > u_{j-1} > u_{j-2}$, i.e. u is monotonic, this parabola is monotonic between x_{j-2} and x_j if

$$\frac{u_{j-1} - u_{j-2}}{u_j - u_{j-1}} \leq 3 \tag{6.72}$$

and is nonmonotonic if

$$\frac{u_{j-1} - u_{j-2}}{u_j - u_{j-1}} > 3. \tag{6.73}$$

Hence the violation of the TVD condition coincides with the nonmonotonicity of the interpolant.

6.8 Total-Variation-Diminishing Schemes with Limiters

The examples in the preceding section suggest an approach to the design of nonlinear TVD schemes with better than first-order accuracy. We saw that first-order backward differencing applied to the linear convection equation with positive a is TVD. When combined with the explicit Euler time-marching method, this spatial discretization will result in numerical solutions with the properties we seek, i.e. no new extrema, nonincreasing maxima, and nondecreasing minima. Hence no spurious oscillations will be introduced into the solution, even at discontinuities. The price paid is that this is a first-order scheme, and the spatial discretization is extremely dissipative. We also saw that second-order backward differencing is not TVD and hence can introduce spurious oscillations. However, it violates the TVD conditions only when the numerical solution has specific properties. This suggests that we can design a nonlinear TVD scheme based on the first-order discretization plus a *limited* amount of a second-order correction as dictated by the TVD conditions.

To illustrate this, we will begin with the second-order backward differencing scheme discussed in the preceding section (6.69). This can be written as the sum of a first-order scheme and a *correction* for second-order accuracy that is simply the difference between the second-order scheme and the first-order scheme. When applied to the linear convection equation, the resulting semi-discrete form is

$$\begin{aligned}\frac{\mathrm{d}u_j}{\mathrm{d}t} &= -\frac{a}{2\Delta x}(3u_j - 4u_{j-1} + u_{j-2}) \\ &= -\frac{a}{\Delta x}[(u_j - u_{j-1}) + \underbrace{\frac{1}{2}(u_j - u_{j-1}) - \frac{1}{2}(u_{j-1} - u_{j-2})}_{\text{for second-order accuracy}}]. \end{aligned} \tag{6.74}$$

This can also be written in conservation form (6.63) with

$$\hat{f}_{j+1/2} = \frac{a}{2}(3u_j - u_{j-1}) \tag{6.75}$$

and similarly split into a first-order term and a correction:

$$\hat{f}_{j+1/2} = a[u_j + \underbrace{\frac{1}{2}(u_j - u_{j-1})}_{\text{for second-order}}]. \tag{6.76}$$

We introduce a limiter ψ to limit the correction as follows

$$\hat{f}_{j+1/2} = a[u_j + \frac{1}{2}\psi_j(u_j - u_{j-1})]. \tag{6.77}$$

If $\psi = 1$, the full second-order scheme is obtained; if $\psi = 0$, the scheme reverts to first-order. Substituting into (6.74) gives

$$\frac{\mathrm{d}u_j}{\mathrm{d}t} = -\frac{a}{\Delta x}[(u_j - u_{j-1}) + \frac{1}{2}\psi_j(u_j - u_{j-1}) - \frac{1}{2}\psi_{j-1}(u_{j-1} - u_{j-2})]. \tag{6.78}$$

From (6.71), the TVD condition for this scheme is dependent on the ratio

$$\frac{u_{j-1} - u_{j-2}}{u_j - u_{j-1}}. \tag{6.79}$$

Hence the value of the limiter function ψ should depend on such ratios. For this purpose, we define the ratio

$$r_j = \frac{u_{j+1} - u_j}{u_j - u_{j-1}} \tag{6.80}$$

and define

$$\psi_j = \psi(r_j) \geq 0. \tag{6.81}$$

With these definitions we can rewrite (6.78) as

$$\frac{du_j}{dt} = -\frac{1}{\Delta x} \underbrace{a\left[1 + \frac{1}{2}\psi(r_j) - \frac{1}{2}\frac{\psi(r_{j-1})}{r_{j-1}}\right]}_{C^+_{j-1/2}} (u_j - u_{j-1}), \tag{6.82}$$

noting that the ratio in (6.79) is $1/r_{j-1}$. Recalling from the TVD conditions in Sect. 6.7 that $C^+_{j-1/2}$ must be nonnegative, we have

$$\frac{\psi(r_{j-1})}{r_{j-1}} - \psi(r_j) \leq 2. \tag{6.83}$$

Given that $\psi(r_j) \geq 0$, the worst case is $\psi(r_j) = 0$, which gives the condition

$$\psi(r_{j-1}) \leq 2r_{j-1}. \tag{6.84}$$

This approach can be repeated for the linear convection equation with a negative value of a and limited second-order forward differencing. One obtains the following symmetry condition on $\psi(r)$:

$$\psi\left(\frac{1}{r}\right) = \frac{\psi(r)}{r}. \tag{6.85}$$

Combining this condition with the above requirement that $\psi(r) \leq 2r$ gives

$$\psi(r) \leq 2. \tag{6.86}$$

Therefore, the requirements that a general limiter function $\psi(r)$ should satisfy include [14]:

$$\begin{aligned}
\psi(r) &\geq 0 \quad \text{for} \quad r \geq 0 \\
\psi(r) &= 0 \quad \text{for} \quad r \leq 0 \\
\psi(r) &\leq 2r \\
\psi(r) &\leq 2 \\
\psi(\tfrac{1}{r}) &= \tfrac{\psi(r)}{r},
\end{aligned} \tag{6.87}$$

where the second requirement arises because a negative value of r indicates an extremum. The value of r can be considered an indicator of the smoothness of the function relative to the mesh spacing. For a differentiable function on a uniform mesh, r approaches unity as the mesh is refined. Hence for second-order accuracy, $\psi(r)$ must satisfy

$$\psi(1) = 1. \tag{6.88}$$

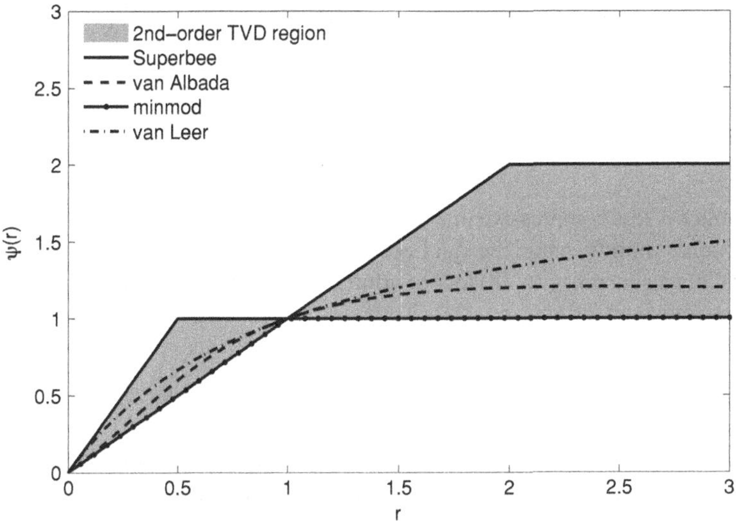

Fig. 6.1 Diagram showing the second-order TVD region for limiters and several well-known limiter functions

Finally, limiter functions lying in the shaded region in Fig. 6.1 lead to second-order TVD schemes [14].

With this flexibility, many different limiter functions have been suggested; some of these are shown in Fig. 6.1. The lower bound of the second-order TVD region is described by the minmod limiter, which is given by

$$\psi = \begin{cases} \min(r, 1) & r > 0 \\ 0 & r \leq 0 \end{cases}. \tag{6.89}$$

Similarly, the superbee limiter [15] lies on the upper bound of the second-order TVD region. It can be written as

$$\psi(r) = \max[0, \ \min(2r, 1), \ \min(r, 2)]. \tag{6.90}$$

The superbee limiter is sometimes referred to as an overcompressive limiter, as it can steepen gradients and introduce staircasing into a smooth solution. This is associated with the fact that $\psi(r) = 2$ for $r \geq 2$. Another disadvantage of both of these limiters is that they are not differentiable for some $r \geq 0$, which can adversely affect convergence to steady state. In order to address this, the van Leer limiter was introduced as follows [16]:

$$\psi(r) = \frac{r + |r|}{1 + r}, \tag{6.91}$$

which is differentiable except at $r = 0$. The van Albada limiter [17] is also differentiable except at $r = 0$:

$$\psi = \begin{cases} \frac{r^2+r}{1+r^2} & r > 0 \\ 0 & r \le 0 \end{cases} . \tag{6.92}$$

This limiter is less compressive than the van Leer limiter as it asymptotes to unity as r approaches infinity, while the van Leer limiter asymptotes to $\psi(r) = 2$.

Next let us generalize (6.78) to a of arbitrary sign. In this case we can define split fluxes as (see Sect. 2.5):

$$f^+ = \frac{1}{2}(a + |a|)u, \qquad f^- = \frac{1}{2}(a - |a|)u. \tag{6.93}$$

With second-order backward differencing for f^+ and second-order forward differencing for f^-, the following limited semi-discrete form is obtained:

$$\begin{aligned} \frac{du_j}{dt} = &-\frac{1}{\Delta x}[(f_j^+ - f_{j-1}^+) + \frac{1}{2}\psi(r_j)(f_j^+ - f_{j-1}^+) - \frac{1}{2}\psi(r_{j-1})(f_{j-1}^+ - f_{j-2}^+)] \\ &- \frac{1}{\Delta x}[(f_{j+1}^- - f_j^-) + \frac{1}{2}\psi\left(\frac{1}{r_j}\right)(f_{j+1}^- - f_j^-) - \frac{1}{2}\psi\left(\frac{1}{r_{j+1}}\right)(f_{j+2}^- - f_{j+1}^-)]. \end{aligned} \tag{6.94}$$

Finally, consider a limited form of the upwind scheme resulting from a piecewise-linear reconstruction, again for the linear convection equation with positive a, as described by (6.45). Writing the numerical flux as a first-order flux plus a limited second-order correction gives

$$\hat{f}_{j+1/2} = au_{j+1/2}^L = au_j + \frac{a}{4}\phi(r_j)(u_{j+1} - u_{j-1}). \tag{6.95}$$

We would like to find $\phi(r)$ such that this is equivalent to (6.77). We obtain the following:

$$\hat{f}_{j+1/2} = a[u_j + \frac{1}{2}\psi(r_j)(u_j - u_{j-1})] \tag{6.96}$$

$$= a\left[u_j + \frac{1}{4}\psi(r_j)\frac{2(u_j - u_{j-1})}{(u_{j+1} - u_{j-1})}(u_{j+1} - u_{j-1})\right] \tag{6.97}$$

$$= a\left[u_j + \frac{1}{4}\psi(r_j)\frac{2}{r_j + 1}(u_{j+1} - u_{j-1})\right]. \tag{6.98}$$

The two forms (6.95) and (6.96) are equivalent if [18]:

$$\phi(r) = \frac{2}{r+1}\psi(r). \tag{6.99}$$

With $\phi(r)$ defined by (6.99), and $\psi(r)$ having the properties described previously, (6.95) is a second-order TVD scheme. In fact it is identical to the scheme defined by (6.77) but written in terms of a piecewise-linear reconstruction. It is easy to show that if $\psi(r)$ has the symmetry property $\psi(1/r) = \psi(r)/r$, then $\phi(1/r) = \phi(r)$.

It can be instructive to write $\phi(r)$ in terms of the undivided differences

$$\Delta_+ = u_{j+1} - u_j, \qquad \Delta_- = u_j - u_{j-1}. \tag{6.100}$$

For example, for $r \geq 0$, the minmod limiter can be written as

$$\phi_j = \frac{2}{\Delta_+ + \Delta_-} \min(\Delta_+, \Delta_-) \tag{6.101}$$

and the van Leer limiter as

$$\phi_j = \frac{4\Delta_+\Delta_-}{(\Delta_+ + \Delta_-)^2}. \tag{6.102}$$

When the limiters are written in this form, their symmetry is clearly evident. It is left as an exercise to the reader to write the superbee and van Albada limiters in terms of Δ_+ and Δ_-.

Furthermore, this form enables a clear understanding of the various limiters and the conditions $\psi(r) \leq 2$ and $\psi(r) \leq 2r$. For example, (6.101) shows that the minmod limiter replaces the slope of the linear reconstruction in (6.95)

$$\frac{u_{j+1} - u_{j-1}}{2\Delta x} \tag{6.103}$$

with the lesser in magnitude of

$$\frac{u_{j+1} - u_j}{\Delta x} \quad \text{and} \quad \frac{u_j - u_{j-1}}{\Delta x}. \tag{6.104}$$

Similarly, the van Leer limiter replaces the slope of the linear reconstruction with

$$\frac{1}{\Delta x}\left(\frac{2\Delta_+\Delta_-}{\Delta_+ + \Delta_-}\right). \tag{6.105}$$

As r tends to infinity, i.e. $\Delta^+ \gg \Delta^-$, this approaches

$$\frac{2(u_j - u_{j-1})}{\Delta x}. \tag{6.106}$$

Similarly, as r tends to zero, i.e. $\Delta^- \gg \Delta^+$, this approaches

$$\frac{2(u_{j+1}-u_j)}{\Delta x}. \tag{6.107}$$

Hence in the limit of large and small r the van Leer limiter produces a slope that is twice that resulting from the minmod limiter. This is a direct consequence of the asymptotic behavior of the corresponding ψ limiters as r tends to infinity, where the minmod ψ limiter is unity for $r \geq 1$ and the van Leer ψ limiter approaches 2 as r tends to infinity. The superbee limiter actually steepens the slope of the reconstruction for some values of r.

Considering the case where u is monotonic and $u_{j-1} \leq u_j \leq u_{j+1}$, in the limit as r tends to zero, the slope

$$\frac{2(u_{j+1}-u_j)}{\Delta x} \tag{6.108}$$

leads to

$$u^L_{j+1/2} = u_{j+1}. \tag{6.109}$$

Thus this is the maximum permissible slope to ensure that $u^L_{j+1/2} \leq u_{j+1}$, as required to preserve monotonicity. Similar arguments apply to $u^R_{j-1/2}$ and to the case where u is monotonically decreasing. Hence the conditions $\psi(r) \leq 2$ and $\psi(r) \leq 2r$ are directly related to the condition that $u^L_{j+1/2}$ and $u^R_{j-1/2}$ lie between u_{j-1} and u_{j+1} unless u_j is an extremum.

The use of limiters to obtain second-order TVD schemes is a powerful and robust approach. There are further issues to be considered, such as convergence difficulties resulting from limiter chatter, systems of equations, multiple dimensions,[3] irregular and unstructured meshes, and higher-order time-marching methods that preserve the TVD property; these are beyond the scope of the present book. Nevertheless, the reader is now in a position to complete Exercise 6.1. The reader can program the second-order flux-split upwind scheme (6.94), a finite-volume scheme based on linear reconstruction (6.95) using the Roe flux function (6.33), or any other second-order TVD scheme found in the literature. The reader is encouraged to experiment with different limiters and other aspects of the algorithm to understand their impact on monotonicity and accuracy. For example, one can reconstruct primitive, conservative, or characteristic variables, and the reader can investigate the impact of this choice on maintaining positivity of pressure and avoiding oscillations.

[3] For example, Goodman and Leveque [19] showed that TVD schemes for scalar conservation laws in two dimensions can be no better than first-order accurate.

6.9 One-Dimensional Examples

As in Chaps. 4 and 5, examples are presented using some of the algorithms described in this chapter. The effectiveness of high-resolution upwind schemes is best demonstrated in the context of the shock-tube problem. The examples provided again coincide with the exercise at the end of the chapter, but the exercise for this particular chapter is more open-ended in that the reader is asked to program the limited second-order TVD scheme of her or his choice. The algorithm demonstrated in the example below is just one of many options.

We consider first a piecewise-constant reconstruction applied in conjunction with the Roe flux function (6.33) and the entropy fix given in (6.39). As a result of the piecewise-constant reconstruction, this spatial discretization is first-order accurate. Time-marching is achieved through the explicit Euler method with a time step of 0.0025 ms. Figure 6.2 displays the numerical solution computed on a mesh with 400 cells in comparison with the exact solution at $t = 6.1$ ms. As expected, the numerical solution is free of unphysical oscillations. However, it also differs quite substantially from the exact solution, in particular near the head of the expansion wave and through the contact surface, which is spread over several cells.

Figure 6.3 displays the numerical solution obtained using the Roe flux function together with a piecewise-linear reconstruction of the conservative variables. This spatial discretization is second-order, and no limiting is applied. Hence the computed solution is more accurate than the first-order solution in some places; however it suffers from large unphysical oscillations and consequently deviates substantially from the exact solution. The results displayed in Figs. 6.2 and 6.3 illustrate the limitations of *linear* schemes in that one can preserve monotonicity of the solution at the expense of accuracy using a first-order scheme or improve accuracy in smooth regions of the flow field using a second-order scheme at the expense of unphysical oscillations and large errors near discontinuities.

Figure 6.4 shows the analogous numerical solution computed using a limited second-order TVD scheme. The slope of the linear reconstruction of the conservative variables is limited using the Van Albada limiter (6.92) following (6.95) and (6.99). The numerical solution is free of unphysical oscillations and is much more accurate than either of the solutions displayed in Figs. 6.2 and 6.3. This solution can be compared with that shown in Fig. 4.18 computed using the less sophisticated shock-capturing approach presented in Chap. 4, which is primarily intended for steady flows (for which it performs very well, as shown in Fig. 4.15). The shock-tube solution computed using the flux-limited scheme is visibly more accurate than that displayed in Fig. 4.18.

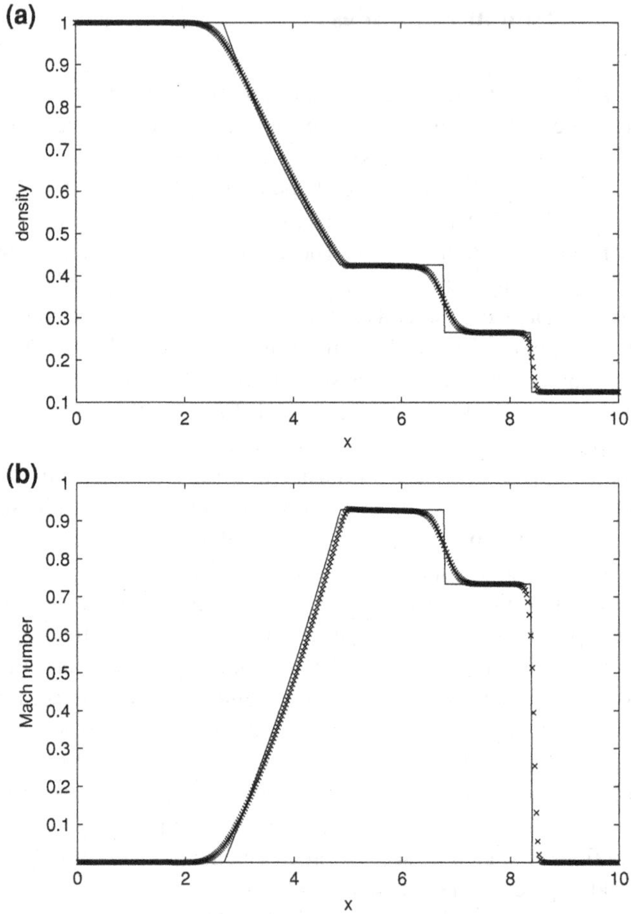

Fig. 6.2 Comparison of the exact solution (–) for the shock-tube problem at $t = 6.1$ ms with a first-order numerical solution based on a piecewise-constant reconstruction (x) computed on a grid with 400 cells with a time step of 0.0025 ms. **a** Density (in kg/m^3). **b** Match number

6.10 Summary

The key topics covered in this chapter include the following:

- Godunov's method: This provides an elegant approach to upwinding for finite-volume methods.
- Roe's approximate Riemann solver and numerical flux function: This provides an inexpensive approximation to Godunov's scheme that satisfies the Rankine-Hugoniot relations at shock waves and is widely used.
- Higher-order reconstruction: Piecewise-polynomial reconstruction enables the development of upwind schemes with spatial orders of accuracy higher than one.

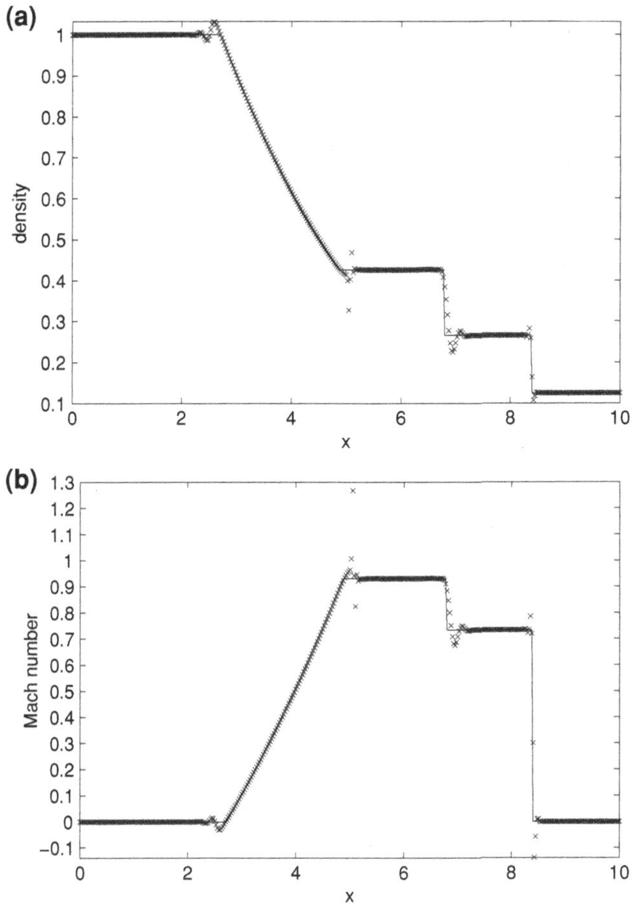

Fig. 6.3 Comparison of the exact solution (–) for the shock-tube problem at $t = 6.1$ ms with a second-order unlimited numerical solution based on a piecewise-linear reconstruction (x) computed on a grid with 400 cells with a time step of 0.0025 ms. **a** Density (in kg/m^3). **b** Match number

- Total variation, monotone schemes, and monotonicity preservation: These provide conditions on numerical schemes such that the resulting numerical solutions have the following important properties: no new extrema are created, local maxima do not increase, and local minima do not decrease.
- Total-variation-diminishing schemes with limiters: These schemes are both second-order accurate and monotonicity preserving. Hence they are very robust, producing oscillation-free solutions that avoid robustness problems such as those associated with negative values of pressure or density.

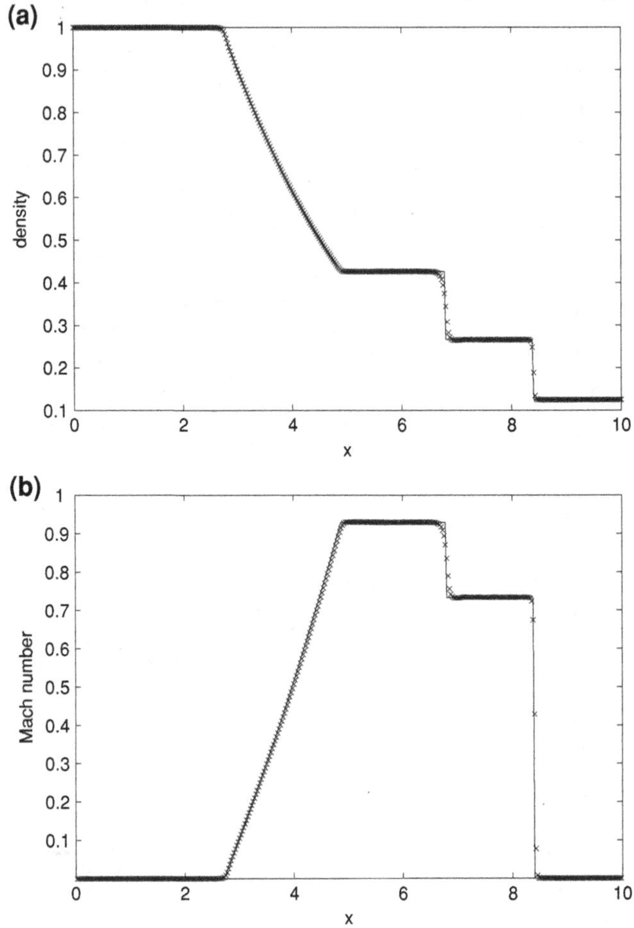

Fig. 6.4 Comparison of the exact solution (–) for the shock-tube problem at $t = 6.1$ ms with a limited second-order numerical solution (x) computed on a grid with 400 cells with a time step of 0.0025 ms. The Van Albada limiter is used. **a** Density (in kg/m^3). **b** Match number

6.11 Exercise

For related discussion, see Sect.6.9.

6.1 Write a computer program to apply a limited second-order TVD scheme to the following shock-tube problem: $p_L = 10^5$, $\rho_L = 1$, $p_R = 10^4$, and $\rho_R = 0.125$, where the pressures are in Pa and the densities in Kg/m^3. The fluid is a perfect gas with $\gamma = 1.4$. Use the explicit Euler time marching method. Compare your solution at $t = 6.1$ ms with that found in Exercise 3.3. Examine the effect of the time step and other parameters on the monotonicity and accuracy of the solution. Compare various limiter functions in terms of their effect on monotonicity and accuracy.

References

1. Steger, J.L., Warming, R.F.: Flux vector splitting of the inviscid gas dynamic equations with applications to finite difference methods. J. Comput. Phys. **40**, 263–293 (1981)
2. Van Leer, B., Flux vector splitting for the Euler equations. In: Proceedings of the 8th international conference on numerical methods in fluid dynamics, Springer-Verlag, Berlin, (1982)
3. Roe, P.L.: Approximate riemann solvers, parameter vectors, and difference schemes. J. Comput. Phys. **43**, 357–372 (1981)
4. Godunov, S.K.: A finite difference method for the numerical computation of discontinuous solutions of the equations of fluid dynamics, Matematicheskii Sbornik **47**, 271–306 (1959)
5. Hirsch, C.: Numerical Computation of Internal and External Flows, vol. 2. Wiley, Chichester (1990)
6. Lomax, H., Pulliam, T.H., Zingg, D.W.: Fundamentals of Computational Fluid Dynamics. Springer, Berlin (2001)
7. Harten, A., Hyman, J.M.: Self adjusting grid methods for one-dimensional hyperbolic conservation laws. J. Comput. Phys. **50**, 235–269 (1983)
8. Van Leer, B.: Towards the ultimate conservative difference scheme. V. A second-order sequel to Godunov's method. J. Comput. Phys. **32**, 101–136 (1979)
9. Lax, P.D.: Hyperbolic Systems of Conservation Laws and the Mathematical Theory of Shock Waves. SIAM, Philadelphia (1973)
10. Harten, A., Hyman, J.M., Lax, P.D.: On finite-difference approximations and entropy conditions for shocks. Commun. Pure Appl. Math. **29**, 297–322 (1976)
11. Lax, P.D., Wendroff, B.: Systems of conservation laws. Commun. Pure Appl. Math. **13**, 217–237 (1960)
12. Harten, A.: High Resolution Schemes for Hyperbolic Conservation Laws. J. Comput. Phys. **49**, 357–393 (1983).
13. Gottlieb, S., Shu, C.-W., Tadmor, E.: Strong stability-preserving high-order time discretization methods. SIAM Rev. **43**, 89–112 (2001)
14. Sweby, P.K.: High resolution schemes using flux limiters for hyperbolic conservation laws. SIAM J Numer. Anal. **21**, 995–1011 (1984)
15. Roe, P.L.: Some contributions to the modelling of discontinuous flows. In: Lectures in Applied Mathematics. vol. 22, pp. 163–193. SIAM, Philadelphia (1985)
16. Van Leer, B.: Towards the ultimate conservative difference scheme. II. Monotonicity and conservation combined in a second order scheme. J. Comput. Phys. **14**, 361–370 (1974)
17. Van Albada, G.D., Van Leer, B., Roberts, W.W.: A comparative study of computational methods in cosmic gas dynamics. Astron. Astrophys. **108**, 76–84 (1982)
18. Spekreijse, S.: Multigrid solution of monotone second-order discretizations of hyperbolic conservation laws. Math. Comput. **49**, 135–155 (1987)
19. Goodman, J.B., Leveque, R.J.: On the accuracy of stable schemes for 2D conservation laws. Math. Comput. **45**, 15–21 (1985)

Index

T. H. Pulliam and D. W. Zingg, *Fundamental Algorithms in Computational Fluid Dynamics*, Scientific Computation, DOI: 10.1007/978-3-319-05053-9,

Printed in the United States
By Bookmasters